THE COLORADO PLATEAU

The

COLORADO
PLATEAU

A GEOLOGIC HISTORY

REVISED AND UPDATED

DONALD L. BAARS

UNIVERSITY OF NEW MEXICO PRESS

ALBUQUERQUE

University of New Mexico Press / Albuquerque

Printed and bound in the United States of America

15 14 13 12 11 10   4 5 6 7 8 9

ISBN-13: 978-0-8263-2301-9

———————————————

Library of Congress Cataloging-in-Publication Data

Baars, Donald L.
        The Colorado Plateau : a geologic history / Donald L. Baars
rev and updated, 1st ed.
        p. cm.
Includes bibliographical reference (p.   ) and index
        ISBN 0-8263-2301-4 (paper : alk. paper)
        1. Geology — Colorado Plateau. I. Title
            QE79.5 .B3 2000
            557.92 — dc21   00–009291

———————————————

DESIGN: *Mina Yamashita*

CONTENTS

PREFACE TO THE REVISED EDITION

The Colorado Plateau is a strange, harsh, but wonderful country, indeed. To many folks, it is a region of precious little vegetation on drab gray plains that stretch endlessly from western Colorado to more lively places like Las Vegas, Nevada or Salt Lake City, Utah. That is because the main highways have been built on the easiest, flattest, smoothest surfaces possible—mostly on exposures of the Mancos Shale. These exposures of thick gray shale are also the sites of many of the largest communities, such as Grand Junction, Durango, Cortez, in western Colorado, for these rocks provide the easiest building sites and the most open valleys. As luck would have it, such vast exposures of the Mancos Shale are also the most difficult construction sites to maintain. Highways sink out of sight, as do airports, and homes crumble or slide down hill with the greatest of ease. Locals call it "gumbo," for good reason.

Yet if one leaves the main towns and thoroughfares and ventures a few miles into the "back country," a fairyland of untold beauty unfolds in almost any direction. There are the many magnificent canyons of the Colorado River system, for which the province owes its name—Grand Canyon, Cataract Canyon, Glen Canyon, and the canyons of the San Juan River to name the more prominent; the colorful monoliths of Monument Valley, Valley of the Gods, Canyon de Chelly; the remarkable fractured tablelands of Canyonlands; the multifaceted rainbow colors of the Painted Desert and Bisti badlands; the hundreds of natural stone sculptures of Arches National Park and the Natural Bridges National Monument; the alpine splendor of the San Juan Mountains in southwestern Colorado. But to enjoy these remarkable scenic wonderlands, one must leave the comfort of primary highways and explore the many less accessible back roads. For those who can break away from the speedways, this is truly an "enchanted wilderness."

What makes the Colorado Plateau unique? It is a broad region of generally simple structural geology, especially as compared to adjacent provinces. Usually shown on a map as an irregularly ovate body of high desert, the Colorado Plateau is in reality a rather stable block of North America that is set off by linear trends in the earth's crust. The major structural trends separate the Colorado Plateau from more chaotic, mountainous regions in all directions. These continental-scale linear structures, sometimes referred to as "lineaments," are complex, deep-seated fault zones that universally trend northeast and northwest, criss-crossing all of the continents, including North America. The fault systems are complicated swarms of faults (fractures along which there has been movement of adjacent rock bodies) that originated more than a billion years ago in Precambrian time. The fault zones are fundamental elements that separate orthogonal blocks of the earth's crust that provide the framework for all younger geologic events.

Segments of the continental-scale lineaments that delineate the Colorado Plateau are shown on Figure 1. As here designated, the Uncompahgre line and a portion of the Rio Grande line separate the Colorado Plateau from the complexly faulted mountain and valley system of the Southern Rocky Mountains on the east. The eastern faulted margin of the Uncompahgre uplift, a highly significant highland and foci of Colorado Plateau stratigraphy throughout much of geologic time, is arbitrarily used to distinguish the two provinces. The Rio Grande rift, extending generally north-south through central New Mexico and northward well into Colorado, might be a more realistic geologic boundary, but that would cut deeply into what geographers call the Rocky Mountains. Anyway, it could be argued that 14,000 foot-high mountain ranges should not be called "plateau" country. We will defy that theorem by including the San Juan Mountains of southwestern Colorado in the Colorado Plateau Province on the basis that the geology in that alpine region has close affinities to Plateau geology.

On the southeast, a relatively obscure line of volcanic and mineralized features, known as the Jemez lineament, can be used to distinguish the topography and geology of the Colorado Plateau from "basin and range" country in southern New Mexico. On the southwest, however, a portion of the Texas-Walker Lane structure, a well-documented fault zone of Precambrian age, clearly separates Plateau country from the fault-block mountain ranges of southern Arizona.

The Wasatch Line on the west is a broad zone that has been a hingeline along which sedimentary rocks thickened drastically westward, perhaps ten-fold, throughout much of geologic time since the Precambrian. The rapid thickening separated the strongly subsiding Cordilleran seaway from the sedimentary platform of the Colorado Plateau for some 350 million years of Paleozoic and early Mesozoic time. Geographically, the line marks the boundary between Plateau country on the east and the "Great Basin," or "Basin and Range" country to the west. The trend is not proven to have originated along a fault zone, mostly because the Precambrian basement occurs at such great depths along the trend as to evade deep drilling operations. The line of great thickening occurs in a linear band in a strongly oriented northeasterly direction, however, as if controlled by typical basement faulting.

The northern border of the Colorado Plateau lies along the southern flank of the Uinta Mountains, a maverick chain that is an enormous east-west range of upfolded (anticlinal) mountains in northern Utah consisting of Late Precambrian sedimentary rocks. The mountain chain lies along an easterly extension of the Mendocino lineament that extends into the Pacific Ocean off the California coast, according to some geologists. The origin of such a strangely oriented mountain range is highly controversial, however, a plausible explanation will be presented in a later chapter of this book.

Many changes have come to light since this book was first published as "Red Rock Country" by the Natural History Press in 1972, and revised as "The Colorado Plateau" by the University of New Mexico Press in 1983. Several pioneers who contributed much to our early knowledge of the geology of the Colorado Plateau have passed away, including friends and mentors such as Edwin D. "Eddie" McKee, Wm. Lee Stokes, Sherman A. "Sherm" Wengerd, Cornelius M. "K" Molenaar and Eugene M. "Gene" Shoemaker. They will be sorely missed, but their valuable contributions to knowledge will live on in the technical literature.

My first wife, Jane, who typed and retyped numerous versions of this book, has also left us for better pastures. My present wife, Renate, has painstakingly prepared many of the illustrations used in this revision.

We have learned many things about the fundamental structure of the earth, thanks to the religious doctrine of plate tectonics, a widely and now almost blindly accepted theory that is based on continents skidding promiscuously across the crust of the earth. Although the doctrine is based on surmised oceanographic processes, all geologic features of the continents must adhere to prescribed policy, even though many of these interpretations make no sense. Field geologists, such as McKee, Stokes, Wengerd, Shoemaker and Molenaar, are no longer necessary, so they say, because we now deal with computer models rather than observed physical facts. If the real geologic facts don't agree with plate tectonics principles, they are simply ignored.

There are a few stalwart individuals who still think the rocks tell us the real story of the earth—who ignore the taunts and obscenities cast on us by plate tectonics disciples. Still, plate tectonics has done some good in that the nonbelievers have been forced to deal with a broader, global perspective of geology, and have had to determine reasonable and realistic alternatives to the prescribed interpretations.

For example, it has become crystal clear to many geologists that "basement structure," the global fault patterns that developed in Middle Precambrian time some 1.7 to 1.6 billion years ago, is of prime importance in localizing and controlling geologic structural patterns on the continents in all of ensuing geologic time. Movement along these "regmatic" fault patterns has been rejuvenated repeatedly throughout time. The existing faults have been preferentially reactivated, rather than remaining passive, with the variations of structural stress fields developed in the earth's crust at different episodes of earth history. In a general way, geologic structures seen at the earth's surface today are the result of a long history of reactivation of fundamental basement structures deep within the earth. The premise upon which this book is based, and the interpretations presented here, are that basement structural rejuvenation has localized and controlled the activity of these ancient fault systems throughout time. There is no need to discuss plate tectonics principles when describing the interior of the continent, as there are no realistic direct relationships to be found.

Another problem of worldwide significance that affects our discussion of the geologic history of the Colorado Plateau, or any other province, is that of the time boundaries of the geologic periods. Most time boundaries have been fairly well established for decades, except the boundaries of the Permian Period. Time boundaries of the Permian have vacillated wildly since the Permian Period was established by Sir Roderick Murchison in 1841 for the region around the city of Perm' in the central Ural Mountains of Russia. Murchison was looking for rocks of an age that would fill the time slot between the Carboniferous System of England (the Mississippian and Pennsylvanian systems as used in North America), and the Triassic System of Germany. The strata he chose in the Ural Mountains only partly filled this time gap, and Russian geologists have argued the exact position of the Carboniferous-Permian boundary for more than 150 years.

Finally, in 1991, Russian geologists sponsored a conference that was held in the beautiful city and countryside around Perm' and proposed a concrete solution to the problem. Details of the proposed changes will be discussed in Chapter three. The practical result of all this is that strata we once called Early Permian (Wolfcampian) are now seen to be Late Pennsylvanian (*Bursumian*—a newly proposed stage name) in age. This may not seem to be important to laymen for, after all, the rocks have not changed, only their assigned ages are different. To the knowledgeable geologist, however, we must be universally correct when applying global terminology to our pet sections of rock. Those folks who enjoy such changes the most are a small group of geologic quacks, mostly retired engineers and Park Service biologists. Regardless of the entertainment value this will bring, the seemingly correct stratigraphic location of the Pennsylvanian-Permian boundary as defined in the type sections in the Ural Mountains has been applied throughout the present revision.

And then, there is the burning question: What constitutes a windblown (eolian) sandstone? There are thousands of cubic miles of sandstone of windblown origin on the Colorado Plateau. Few geologists doubt that the Coconino Sandstone of northern Arizona consists of "fossil" sand dunes, or that the widespread Navajo Sandstone was a vast Sahara-like desert in Jurassic time. But what about all those other sandstones: The White Rim?; the Cedar Mesa?; the Wingate? And where did all that sand come from, anyway? In the olden days, back when I was starting my career, any sandstone with large scale cross bedding was considered to be windblown, or "eolian." Cross bedding consists of bedding planes that slope at some angle to the horizontal. The condition forms where sand grains are rolled up the windward slope of a dune by current action, and then settle on the leeside of the dune where the current slackens. The process forms deposits at the natural angle of repose—that is, at the highest angle at which sand can accumulate without slumping. If that slope becomes oversteepened, slumps or avalanches will form that will become the bedding plane. The usual angle of repose for sand deposits formed in air is about 32 to 35 degrees from the horizontal; that for underwater dunes is significantly lower in angle. Classic examples of such wind-blown cross bedded sandstones are seen in the Coconino Sandstone of Grand Canyon and in the Navajo Sandstone, especially in the Zion National Park region of southwestern Utah.

A great deal of research effort has been spent on studying the internal characteristics of windblown deposits. Eddie McKee, the late grand daddy of Grand Canyon geology, became interested in the subject many years ago and began digging trenches across sand dunes all over the world. His technique was simple. He would soak a dune with water, then dig or bulldoze vertical trenches at various angles through the dune. Internal bedding features would thus be exposed for photography and sampling. When McKee's technical papers became widely read, numerous others thought that it would be great fun to run around the world digging in the sand. It was in this way that we now know that dune sands can accumulate in every arrangement known to man, and a few others as well.

Meanwhile, everyone agrees that sand is carried and deposited by water currents, yet no one has found a way to dig trenches that will stand still in oceanic settings. One cannot find a single reliable technical paper on what is to be found inside water-deposited dune sands. Externally, oceanic sand accumulations look and act like windblown sands, forming sand bars of various shapes and even the arcuate "barchan" dunes so typical of windblown deserts; ripples of all kinds form commonly. The internal characteristics remain unknown except for studies conducted by McKee and a few others in wave tanks in the laboratory.

So we end up knowing everything there is to possibly know about windblown sand deposits, and virtually nothing about water-laid sand accumulations. Therefore, there can be no argument concerning the origin of any sandstone. The windblown-sand worshippers rule the world by the volume of published data. All cross-bedded sandstone is of eolian origin—or else!

As we will see during the course of this discussion, I am neither a worshipper of plate tectonics, nor do I believe that all sandstone is of windblown origin. Questions still must be asked! The correct or best answers must still be sought! Hopefully, "the truth will out" in the long run.

Three other books have been published recently that update the geology in some detail in specific regions of the Colorado Plateau. These are: *Canyonlands Country*, published by the University of Utah Press, *The American Alps* (San Juan Mountains), and *Navajo Country*, both published by the University of New Mexico Press. The original version of this book, *Red Rock Country* and later *The Colorado Plateau—A Geologic History*, were attempts to bring the fascinating story of the geology of the Colorado Plateau to the attention of interested nongeologists. It seems that the effort had mixed results, being perhaps too technical for some readers but useful to students. This revision will attempt to further simplify the presentation to appeal to readers at all levels, and to bring together and perhaps generalize the stories told in the more restricted regional books mentioned above. In other words, the effort has come full circle, as has the career of the author, hopefully to be more meaningful in this last version. Perhaps this generalized account will spur some young geologists to come down from the plate tectonics conveyor belt, to relegate computer models to weekend party games, and get back to the field where they may find answers to many of the questions raised here. More questions remain than have been answered—the next round of answers will come from diligent field work, not from arm-waving visionaries.

PART ONE
The Early History

Angular unconformity in Box Canyon near
Ouray, Colorado. Lower vertical beds are in
the Late Precambrian Uncompahgre Formation;
near horizontal beds above are in the Late
Devonian Elbert Formation. The Precambrian
quartzite layers were deposited in a horizontal
attitude, then folded to nearly vertical and
eroded to a plain. The upper beds were
deposited on the upturned and eroded seafloor.

CHAPTER ONE

THE GRAND CANYON

AN OPEN BOOK

The Grand Canyon of the Colorado River is one of the most magnificent natural spectacles of this earth. Its mile-deep chasms and intricate tributary gorges provide the greatest scenery in the United States and an inspiring recreational area for people of all ages. To those who view it from the rim, it is an unbelievably large gash into the otherwise nearly flat plateau that surrounds it in all directions. Its walls are obviously composed of numerous layers of multicolored rocks that appear to change colors as readily as a chameleon with the changing light of evening. This, to most visitors, is the magic of Grand Canyon.

To those hardier souls who venture into the canyon, the huge size and grandeur of Grand Canyon take on greater meaning. The rock layers become formidable cliffs that are insurmountable except by way of narrow, winding trails that have been carved from the precipice. The tributary gullies that seemed insignificant from above become major canyons in their own right, providing myriad obstacles to travel along the layers. As one descends to the inner realms of the canyon, the walls and remnant towers of the sinuous tributaries become towering mountains that guard the river from intruders. The great depth of the canyon is better appreciated when it is realized that one is passing a continuous series of plant and animal habitats that generally range from southern Canada to the deserts of Mexico; average temperature variation is about 20 degrees from the canyon rim to the Colorado River. The size of the canyon becomes even more profound to the hiker attempting to climb back out of the hole he has gotten himself into.

The scenic grandeur of the canyon becomes insignificant and superficial when compared with the exciting tales that are written in the rocks of the grandest of canyons. The tireless and timeless work of the Colorado River with its relentless tools, the sand and gravel it carries, has carved this abyss through thousands of feet of rock layers that represent hundreds of millions of years of geologic time and untold generations of living beings, back to the earlier forms of life, some two billion years old. The rapid erosion of the canyon in this arid climate has produced remarkably well-exposed views of the rock layers, which tell fascinating stories to anyone who is willing to take the time and effort to read it. The casual observer will probably notice sea shells in the limestones that cap the canyon rim, representing organisms that lived in a warm tropical sea that covered the region approximately 250 million years ago. With every step one takes downward into the canyon, perhaps a thousand years of earth history are reviewed and older and more primitive fossils may be found preserved in the rocks. Finally as one enters Granite Gorge, in the inner depths of the canyon, one moves over rocks that were formed before there was any kind of life on earth, except primitive bacteria and algae, some two billion years ago. Here, in Grand Canyon, where layers are exposed like pages in a book, is a fitting place to begin our journey back into the beginnings of life and time itself.

The Oldest Rocks

The oldest rocks on earth that have been dated are more than four billion years old. The ancient rocks in the depths of Grand Canyon are not that old; however, it is certain that they are older than the earliest time of abundant life on earth, which geologists call the Cambrian Period. The rocks in the inner gorges were formed during the first great era of earth history, known as the Precambrian Era, which began when the earth was formed, and continued until about 570 million years ago. This era lasted some 3.5 billion years. This long period of time is lumped together into one era because the very old rocks have undergone such a long history of burial at great depths and such intense deformation that they are very difficult to study and date. With the advent of life, capable of being preserved in the rocks as fossils, the rocks could be dated much more closely and accurately. However, Precambrian rocks must be studied without the aid of fossils; only the relationships between rocks can be used as a guide to their antiquity. Although the ancient rocks can now be dated by measuring the rate of disintegration of radioactive minerals within them, few have actually been dated in this way because of the rather large expense involved and the many uncertainties that accompany the interpretation of the measurements.

Grand Canyon from the south rim. The entire Paleozoic section is exposed in the alternating ledges and slopes from the erosive processes of the Colorado River as it carved its path into the nearly flat-topped plateau.

The rocks that were formed during Precambrian times are usually very twisted and distorted because of the effects of the great temperatures and pressures to which they have been submitted during their long history. Such rocks are called "metamorphic rocks" because of their changed appearance. Most of the rocks in the bottom of Grand Canyon have been highly altered and completely recrystallized, so their original appearance is destroyed and they now appear as dark-colored platy and banded rocks known as "gneiss" and "schist." These rocks, now called the "Vishnu Schist," were originally formed from sand and mud deposited in an ancient primeval sea. Later, they were folded and broken to form great mountainous massifs. Molten fluids were squeezed into the highly folded rocks at considerable depths to form the pink or cream-colored bands of granite that punctuate the cliffs of the inner gorges, filling the original fractures. The greatly altered condition of the rocks of the inner gorges suggest that several episodes of mountain building may have occurred before the forces of erosion reduced the mountains to a level plain. While this may sound exciting and catastrophic, it probably took hundreds of millions of years to accomplish and probably was no more noticeable than the crustal movements and erosion that are occurring around us today. The Vishnu Schist as we now see it in the deep inner gorges of Grand Canyon represents the roots of an ancient mountain system, probably much like the Himalayas of today, that was reduced over great lengths of time to a nearly smooth plain by painfully slow erosion.

No recognizable traces of life have been found in the Vishnu Schist, although very primitive forms of life may have already been evolving when it was forming. The fossilized remains of primitive algae, simple plants similar to modern algae that form scum on ponds and other damp surfaces, have been found in south central Canada in rocks about 2 billion years old. Similar plants or bacteria may have been present when the Vishnu Schist was forming, but if such organisms were present in the original sediments they have been completely destroyed by the intense metamorphism.

The Vishnu Schist may be thought of as the "basement rocks" upon which the remainder of the geologic record was constructed. The oldest rocks invariably lie at the bottom of a sequence of rock layers, unless, as is rarely the case, extreme folding and overturning of the earth's crust has occurred. This basic principle of geology, based on the premise that the older rocks must have been present before the younger rocks could have been deposited on them, is called the "law of superposition." Because of it, a geologic section such as the one exposed in Grand Canyon is better studied from the bottom up, since this was the order in which it was deposited.

Precambrian Sedimentary Rocks

When the Vishnu Schist had been worn down to a nearly level plain by erosion, the sea again moved gradually into the area now occupied by Grand Canyon. The surface of erosion, which in Grand Canyon is a very abrupt change from dark metamorphic rocks to gray or red sedimentary rocks above, is called an unconformity. This general term implies a break in the depositional history (or a break in the normal "conformity" of geological events), and may be used for any major interruption in the normal deposition of sediments. Sediments of sand, mud, and lime were deposited on the sea floor directly on top of the old eroded surface. When compacted and cemented together, these sediments became sandstone, shale, and limestone respectively.

The sedimentary rocks that were deposited on the unconformity are called the Grand Canyon Supergroup. They consist of some 13,000 feet of sandstone, shale, limestone, and rock made up of cemented pebbles called conglomerate, with a few lava flows interspersed. This very thick deposit of sediments, like the Vishnu Schist, was deposited prior to Cambrian time, and consequently included in the Precambrian Era. That it is much younger than the underlying metamorphic rocks is demonstrated by the fact that the Grand Canyon Supergroup has not been subjected to the process of intense metamorphism that altered the Vishnu rocks. In other words, the metamorphic process that altered the older rocks to gneisses and schists were completed before the younger Grand Canyon Supergroup was deposited.

All was not peace and serenity for the younger Precambrian rocks, however, for shortly after the sediments were compacted to form rocks, they were broken by long fractures, or faults, in the earth's crust and then were tilted into a series of elongate uplifted ridges. These fault blocks trend in a northwesterly direction and tilt the strata towards the northeast. When the seas withdrew and sedimentation ended in the region at the end of Precambrian time, this faulted surface was subjected to the destructive forces of erosion and was partially worn down. The harder rocks, such as the hardest sandstones and the limestones, stood in relief above the surrounding terrain, forming hogbacks, but the softer rocks such as the shales were eroded into smooth slopes and elongate valleys between the fault blocks. Such hogbacks are forming commonly today wherever tilted beds of rocks at the earth's surface are exposed to the processes of weathering.

When the sea again encroached upon the land at the beginning of Cambrian time, this weathered surface was covered by new sand deposits surrounding the hogbacks, which then stood as islands in the shallow sea. This process of tilting the older rocks, then subjecting them to erosion, and finally covering them with flat-lying younger sediments produced an unconformity where horizontal rocks cover the upturned edges of the eroded older rocks to form an "angular unconformity." Tilted strata of the Grand Canyon Supergroup and the prominent angular unconformity can be seen from many vantage points along the rims of eastern Grand Canyon.

The early traces of life on earth appear in rocks of the Grand Canyon Supergroup. The shales and limestones contain minute fragments of shell-like

ERA	PERIOD	APPROX. AGE IN MILLIONS OF YEARS	APPROX. DURATION IN MILLIONS OF YEARS	GRAND CANYON
Cenozoic	Quaternary	0-1	1	Vulcan's Throne volcanics
	Tertiary	1-65	64	Canyon erosion
Mesozoic	Cretaceous	65-135	70	
	Jurassic	135-195	60	
	Triassic	195-230	35	
Paleozoic	Permian	230-290	60	Kaibab Fm.
				Toroweap Fm.
				Cococino Ss.
				Hermit Sh.
				Esplanade Ss.
	Pennsylvanian	290-325	35	Wescogame Fm.
				Manakacha Fm.
				Watahomigi Fm.
	Mississippian	325-355	30	Redwall Ls.
	Devonian	355-410	55	Temple Butte Fm.
	Silurian	410-440	30	(none preserved)
	Ordovician	440-500	60	(none preserved)
	Cambrian	500-570	70	Muav Ls.
				Bright Angel Sh.
				Tapeats Ss.
Precambrian	Younger	570-1,700±		Grand Canyon Super Group
	Older	1,700-4,500±		Vischnu Schist

The Esplanade Ss., Wescogame Fm., Manakacha Fm., and Watahomigi Fm. are grouped as the Supai Group.

Geological abbreviations: Fm. = Formation Ss. = Sandstone
Sh. = Shale Ls. = Limestone

material that may have been formed by primitive sea-dwelling organisms. The imprints of regularly shaped structures that have been interpreted as molds of ancient jellyfish have been found in these sedimentary layers. While these structures and the shell-like fragments are of problematic origin, plant spores called *Chuaria* have been identified in the upper beds of the sequence. Delicately laminated rocks called stromatolites, which are formed by the life processes of primitive algae, are common in some of the limestones. Similar 2 billion year old stromatolites in Ontario, Canada, were replaced almost immediately by chert, a fine-textured, dense siliceous rock, which preserved the remains of one-celled and filamentous algae in unbelievable detail.

To better understand both how the microscopic algae can produce these laminated stromatolites and their significance, let's take a quick look at a modern environment where similar deposits are being formed today. On many of the low-lying islands in Florida Bay off the southern tip of Florida and in the Bahama Islands to the east, the intertidal mud flats are covered by a scum or mat of living microscopic blue-green algae. These primitive plants live on the damp surface in this tropical climate and make their own food by utilizing the carbon dioxide in the air and sunlight to form carbohydrates by photosynthesis. When a particularly high tide or storm occasionally brings a fresh supply of sediments onto the mud flats, the algal mat is buried, and the life-giving photosynthetic processes come to a halt until the algae can wriggle up through the thin layer of mud or silt to again become established on the upper, sunlit surface. The newly established organic layer on the sediment surface traps and binds down the fine-grained sediments so that they are not easily removed by erosion. This forms a single lamina of sediments. When the next layer of sediments is brought in on the high tide and trapped and bound down by the algal mat, another lamina is produced, and so an. If similar processes formed stromatolites in Precambrian time (and there is no reason to think otherwise), these laminated limestones of the Grand Canyon Supergroup were probably deposited on ancient tidal flats in a tropical climate.

The idea that processes acting on the earth's surface today are similar to the processes that formed the features we see in the ancient deposits is the basis of interpreting the geologic record. It is the tool used to determine the origin of the sedimentary particles, the unusual markings that are often present in the rocks, the nature of the environment of deposition, and even the climate in which the sediments were deposited. Modern sites of deposition are studied to learn what process produces what sedimentary features, and then the ideas so developed are used to interpret the rocks. This process is based on the "doctrine of uniformitarianism," which states simply: "The present is the key to the past."

The Geologic Time Scale

Before continuing our journey through the ancient history of Grand Canyon, it would be well to stop briefly to examine the manner in which geologists divide geologic time into useful and understandable units. The main intervals of geologic time are the eras and the periods, which are distinguished in order to classify the millions of years that have passed during the formation of the earth as we know it. The need to subdivide time is primarily based on our inability to comprehend such great intervals of time and great thicknesses of strata, and also because it is so difficult with the presently available geologic tools to accurately assign absolute dates in years to geologic events. Therefore, segments of time have been set aside for our aid in reconstructing the events of the past into a coherent story.

All of geologic time is subdivided into large units called eras; the names assigned to them are based on the stage of development of their contained fossils. The name of the latest era, the Cenozoic, comes from the Greek words kainos, or "recent," and zoe, or "life": "Recent life." Similarly the next-older era, the Mesozoic, refers to "middle life," and the older Paleozoic Era signifies "ancient life." The Precambrian was originally divided into two eras: the younger Proterozoic, or "earliest life," and the older Archaeozoic, or "initial life." Because of the difficulties of dating these ancient rocks, they are often now referred to by the more general term Precambrian for simplicity. The eras are shown on the left side of the time scale (page 8).

The eras are subdivided into smaller, more usable intervals called periods. These units were originally named for areas in the world where rocks of the age in question are well developed and well exposed. Periods are the fundamental units of the geologic time scale. The periods are shown on the time scale in their proper sequence, with the approximate portion of absolute time that they represent shown in years. An example of how the periods were originally defined is shown by the history of the Cambrian Period. This period was named by Adam Sedgwick in 1835 for a series of about twelve thousand feet of very fossiliferous sedimentary rocks from Cambria, the Latin name for Wales. Now, strata throughout the world that are believed to have been deposited at about the same time as these strata in Wales are said to be of "Cambrian age." Although the Cambrian Period is believed to have occurred between 500 and 570 million years ago, it is usually impossible to determine the absolute date in years for rocks deposited during this period, so they are usually dated in a relative fashion by the fossils they contain.

The fossils can be used as tools for dating rock strata because life has continuously evolved from the simplest forms toward more-complex organisms. This evolutionary process is shown by the successive changes in fossils in progressively younger rocks in an orderly fashion, so that once the proper succession of the fossils is determined in known areas, they can be used for dating unknown rocks. This is done by the identification of the fossils, so that they may be placed in their proper horizon of evolutionary development and then simply matched up with the fossils from a known age. Thus, the state of evolutionary development is the tool that enables the determination of relative ages of the strata.

Fossils are useful for purposes other than dating the rocks. It is well known that modern plants and animals inhabit only particularly well-suited environments in considerable numbers, and do not live at all in exceptionally unfavorable habitats. For extreme examples, whales do not live in freshwater lakes and dogs do not dwell in the ocean. Consequently, the environment of deposition of the sedimentary rocks can be determined to some extent by the kinds of fossil organisms they contain, if we ignore any fossils that were carried into the area from elsewhere after death. Such things as water depth, temperature, amount of wave action, water salinity, and other such characteristics can often be determined by studying the fossil content of the rocks. An example of determining the environment of the formation of stromatolites has already been cited.

The formal periods of geologic time that are represented by the rocks of Grand Canyon are shown at the right of the time scale. All the rocks that lie above the Grand Canyon Supergroup within the canyon proper were deposited during the Cambrian, Devonian, Mississippian, Pennsylvanian, and Permian Periods. Rocks representing the other periods are present in nearby regions, but for one reason or another, which will be explained during the course of this discussion, were not preserved in Grand Canyon. It would be desirable to look over the time scale and become familiar with the terms and their order of succession, for they will be used throughout the remainder of this book.

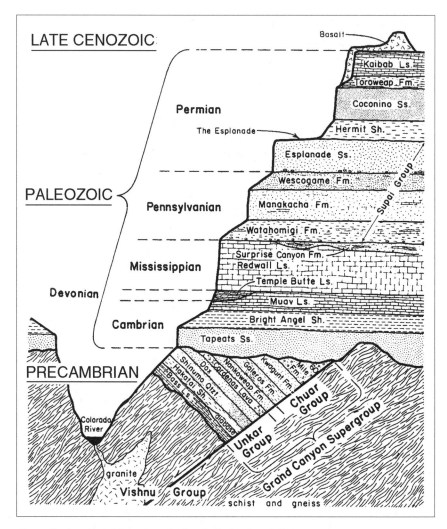

LATE CENOZOIC

Basalt
Kaibab Ls.
Toroweap Fm.
Coconino Ss.

PALEOZOIC

Permian
The Esplanade
Hermit Sh.
Esplanade Ss.
Wescogame Fm.
Manakacha Fm.
Watahomigi Fm.
Supai Group

Pennsylvanian

Mississippian
Surprise Canyon Fm.
Redwall Ls.
Temple Butte Ls.
Muav Ls.
Bright Angel Sh.
Tapeats Ss.

Devonian

Cambrian

PRECAMBRIAN

Shinumo Qtzi.
Hakatai Sh.
Bass Ls.
diabase
Dox
Cardenas Lava
Nankoweap Fm.
Galeros Fm.
Kwagunt Fm.
Sixty Mile

Unkar Group
Chuar Group
Grand Canyon Supergroup

Colorado River
granite
Vishnu Group
schist and gneiss

*Generalized stratigraphic section of rock units in the Grand Canyon.
From Potochnik and Reynolds, 1990.*

The Dawn of Life—Cambrian Period

The Paleozoic Era dawned in the Grand Canyon country when the sea migrated slowly eastward over the upturned and eroded edges of the Precambrian terrain. The initial sediments of this period were deposited along the shoreline in the form of beaches and related sand dunes and sand bars. The sand must have come from the weathering of the Precambrian rocks that formed the country rock along the shore. The beach environment at the shoreline of this great sea probably extended for many miles in a general north-south direction, but was only a few yards wide, as beaches of today are. As the sea marched across the northern Arizona lowlands, a result of a general relative rise in sea level in the great Cordilleran sea of Nevada, western Utah, and northward into western Canada, the shoreline and its close relative the beach moved gradually eastward. This migration of the shoreline caused the deposits of the narrow beach to spread out over all of northern Arizona and eastern Utah, forming the Tapeats Sandstone. Because the eastward march of the beach was slow and gradual, the sandstone was deposited at an earlier time in the west than farther to the east, where the shoreline did not reach until considerably later. Thus, an extensive sheet of sand was deposited as a result of a transgressive shoreline, being of Early Cambrian age in the west and gradually climbing in time until it was deposited in the Late Cambrian in the east.

Meanwhile, mud was being deposited farther from shore in the deeper, quieter waters. Although these muds were being deposited seaward from the beach, they gradually came to overlie the beach sands as the shoreline and the offshore mud area migrated together in an easterly direction. These marine muds later became lithified by compaction to form shale layers, which are called the Bright Angel Shale for their thick and well-exposed outcrops along Bright Angel Creek in Grand Canyon. Such sequences of uniform rock type as the Bright Angel Shale and the Tapeats Sandstone that may extend for many miles in all directions are formally called "formations" and are named for areas in which they are well developed. This area, which is chosen because it is a very good place to go if you want to study the formation, is called the "type section."

Still farther offshore from the sand and mud areas, lime muds composed largely of sediments derived from the disintegration of the hard shelly parts of plants and animals were deposited in relatively deep waters. These fine limy sediments accumulated westward (seaward) from the sand and mud areas, but as the sea marched toward the east they, too, came to overlie the muds. This limestone formation, known as the Muav Limestone, was slowly replaced upward by dolomite, limestone that is altered by the chemical addition of magnesium, that has never been given an official name. Because of extensive stromatolite occurrences in the unnamed dolomite layer, it is interpreted to have been deposited in lime-rich intertidal mud flats as the Late Cambrian sea withdrew westward back to the Cordilleran seaway of

Nevada and western Utah. The unnamed formation was the last and youngest deposit of the Cambrian sea and forms the top of the Cambrian sequence.

The four formations of the Cambrian form the broad Tonto Platform of the lower Grand Canyon, which extends the entire length of the canyon. The lower Tapeats Sandstone forms the steep brown cliffs that directly overlie the Vishnu Schist of Granite Gorge, in the western part of the canyon, and rest on the upturned edges of the Grand Canyon Supergroup in the eastern parts of the canyon. The Tapeats is the lowermost of the horizontal strata that overlie Precambrian rocks in Grand Canyon; the base of the Paleozoic rock sequence. The Tapeats grades upward into the drab-colored shales of the Bright Angel, which, because of their soft and easily eroded nature, form the broad, gentle slopes of the Tonto Platform. The shale bench is capped by the abrupt cliffs of the Muav Limestone and unnamed dolomite. Because of their close relationship to one another and their prominent display along the Tonto Platform, the Cambrian formations are combined into a four-part unit called the Tonto Group.

Fossils are fairly common in the Cambrian strata, marking the earliest time when organisms that were easily preserved as fossils came into existence. The most abundant and most interesting of the Cambrian fossils are small organisms called trilobites. What have been fossilized are the protective hard parts that grew on the backs of the crablike animals. In spite of the fact that trilobites are among the oldest fossils ever found, they were quite advanced, as invertebrate marine organisms go. They were segmented much as are modern worms, but they also moved on jointed limbs like those of insects. The body was divided into three basic parts, which may be thought of as a head, a thorax or body area, and a tail region. The hard shell provided protection to their backs, which were exposed as these now-extinct organisms crawled or swam along the sea floor. The trilobites, which vary from a fraction of an inch to three inches in length in the Tonto rocks, evolved very rapidly with time and so are excellent fossils to use in dating the Cambrian strata. It was through careful studies of the trilobites that the Tonto formations were first found to be considerably older in the west. Older trilobites occur closer to the base of the Tonto sequence toward the west, finally being replaced by higher and consequently younger species in the east.

A Break in The Record

There is no record of rocks of the succeeding two geologic periods, the Ordovician and Silurian times, having been deposited in the Grand Canyon region. The next strata above the Tonto Group are of Devonian age, demonstrating that there was a significant amount of time, perhaps 100 million years, during which no rocks were preserved. It may be that no rocks were ever deposited during this time and that the region was slightly elevated above sea level, where neither erosion nor deposition could occur. Another possibility is that some sediments were deposited as thin layers across this region and were later removed by erosion prior to deposition of the overlying Devonian beds. At any rate, an unconformity representing a hiatus of considerable duration separates the Cambrian rocks from the next-younger beds. That there was an interval of time of this duration at all is shown by several thousand feet of strata of Ordovician and Silurian age preserved in the western Utah and Nevada Cordilleran seaway to the west and northwest of Grand Canyon. Consequently a great skip in the evolutionary traces of life occurs at this horizon in this region.

The Age of Fishes (Devonian Period)

The processes of erosion during the Ordovician and Silurian periods left a relatively smooth surface on the Cambrian strata with a few stream channels cut into it. Sands and lime sediments accumulated on this surface in Devonian times, but were not preserved over much of eastern Grand Canyon region due to later erosion. The Devonian rocks, called the Temple Butte Formation, are preserved only in the small channels at scattered localities where pre-Mississippian erosion did not completely remove them. One of these remnants is visible along the Kaibab Trail at the top of the Cambrian strata, and several more are visible in the walls of Marble Canyon upstream from the mouth of the Little Colorado River. Although occurrences of Devonian rocks are rare within eastern Grand Canyon, they thicken westward to a thickness of about 1,200 feet in the western canyon, thickening toward Nevada and the Cordilleran seaway. Devonian strata are well known in other localities toward the north, northeast, east and south of the canyon proper. This small area of largely missing Devonian rocks probably resulted from the region of eastern Grand Canyon having been somewhat higher than the surrounding regions either during sedimentation (so that the sediments did not accumulate to such thicknesses) or shortly after deposition (so that more strata were removed by erosion in the vicinity of Grand Canyon). In either case, it is probable that the distribution of the Temple Butte Formation reflects an early episode of gentle elevation along the eastern Kaibab uplift, a broad region of arched rocks through which the Grand Canyon was cut in much later geologic times.

The Devonian Period was a time when the highest form of life on earth was the fishes, and most of these were primitive creatures that carried a heavy coat of bony armor for protection from their enemies. Fragments of the plates from these fish, which are believed to have been of fresh-water habitat, have been found in the Temple Butte Formation. The fish remains have been studied and compared with similar fossils elsewhere, and thus have been used to date the Temple Butte Formation as Late Devonian in age. If the fish were truly fresh-water creatures, then the Devonian deposits in the Grand Canyon region are either of fresh-water origin or they represent marine deposits with a nearby source of fresh-water materials such as may have been supplied by a river. The latter is probably the most logical explanation.

The Great Tropical Sea (Mississippian Period)

One of the most prominent single layers in Grand Canyon is a massive reddish cliff of limestone that occurs about halfway up the walls of the canyon. This cliff can be traced for the entire length of the canyon as a geologic bookmark, which separates the older Paleozoic rocks from the Permian strata. It is called the Redwall Limestone because of its characteristic reddish color on the weathered outcrop. The limestone itself is gray when seen on a freshly broken surface, and its reddish color is found to be only skin deep. The red iron-oxide paint was supplied from the red shales of the overlying Permian strata, which wash down from above during flash floods and spring runoff. The cliff is between five hundred and eight hundred feet high in most places, directly reflecting the thickness of the formation, which weathers to a prominent vertical-to-overhanging cliff in this relatively arid climate. As such, the Redwall Limestone is a major obstacle to travel into the canyon, for it must be crossed, in most cases, where trails have literally been carved from the rocky cliffs.

The Redwall Limestone in Grand Canyon is part of a very extensive blanket of limestone that was deposited all along the eastern shelf of the Cordilleran sea in the early half of the Mississippian Period. It is known throughout the Rocky Mountains and the Colorado Plateau country, extending from Mexico into Canada. It is no wonder, then, that it can be seen extending the full length of Grand Canyon. It is usually a fossiliferous gray limestone, although dolomite may be very common as result of alteration of the original limestone by the addition of magnesium and a related recrystallization of the original texture. Chert (concentrations of dense siliceous rock in thin beds and nodules) is usually present, especially in the lower half of the formation.

The sediments that constitute most limestones come from the shells and other limy hard parts of plants and animals that live in tropical seas. Fossils and fossil fragments often make up most or all of the rock, but other limy grains are very common constituents. Many of the lime sediments are derived from the breakdown of larger shell particles by scavenging organisms that bore into the shelly material or fragment it by swallowing larger particles and reducing the grain size by physical force or organic acids as it passes through their bodies. The nondigestible, limy parts are then excreted in the form of fine sediments or in pellets composed of compacted particles of the fine-grained calcium carbonate. Much of the fine-grained lime sediment is probably produced in this way, although many tropical marine algae form mud-sized particles of calcium carbonate as a by-product of their photosynthetic processes. These particles become lime mud when the alga dies and disintegrates.

Small, round grains of calcium carbonate called oolites are also common in the Redwall and its related formations. These small sand grains result from the direct precipitation of calcium carbonate from supersaturated seawater onto preexisting sand grains that act as nuclei around which the limy layers grow. The resulting grain has a nucleus of a sand particle surrounded by concentric shells or laminae of very small crystals of calcium carbonate.

Any or all of these kinds of lime sediments may be segregated into sands or muds by the waves, tides, and other oceanic currents, and deposited wherever the currents slacken so that the sediments can no longer be maintained in suspension. Although the sediments are manufactured locally by the organisms that live there, they are otherwise similar to other sands and muds and may be transported and deposited in much the same fashion. Organic reefs and banks constructed on the spot by organisms such as corals differ from other kinds of sedimentary deposits in that they are produced where they are found and do not depend on transportation and deposition for their occurrence. Because most plants and animals that produce limy hard parts live in shallow tropical seas, most fossiliferous limestones are believed to be of tropical marine origin. The lime sediments become rock by their cementation with lime cements.

The Redwall Limestone is commonly fossiliferous, containing the shells of marine organisms mixed in with other kinds of lime sediments. Among the most common fossils in these rocks are the remains of crinoids, or sea lilies, which were animals that lived attached to the sea floor by a stemlike structure in much the same manner as plants. The organism lived primarily in a calyx, or headlike structure, from which arms projected for the gathering of food. The calyx was protected by plates of calcium carbonate, and the stemlike column was composed of a chain of calcium-carbonate buttonlike structures that, when attached end to end by organic matter, formed the stalk. These hard lime parts are very common constituents of these limestones. Another common fossil in the Redwall and other Paleozoic formations is brachiopod shells. Brachiopods, or lamp shells, are marine bivalves similar in general appearance to the more common clams in modern seas. They live attached to the sea floor by fleshy stalks and feed by filtering microscopic organisms from the sea water as it passes them by. Brachiopods are fairly large fossils, ranging up to about three inches in length, and are often used to date the rocks because most of them evolved quite rapidly. Other common fossils in these rocks are solitary corals, which look like small "horns of plenty," made of a calcium carbonate shell in which the coral polyp rested in the sediments. These did not form massive coral colonies or reefs in this region, but are quite useful as age indicators.

One of the best groups of fossils for dating the Redwall and other formations of Mississippian age are the shells of microscopic single-celled organisms called Foraminifera, or "forams" for short. This group of organisms as a whole lived throughout most of geologic time since the Cambrian, but one group known as the endothyrid forams were particularly plentiful and are useful in dating Mississippian rocks. These tiny creatures drifted about with the tides or lived along the bottom of the sea, evolving so rapidly that rocks deposited during very short intervals of Mississippian time can be recognized and dated if the forams can be found. Because they are so small, they are usually found and studied in thin-sections, which are slices of rock cut so thin that light can pass through them. These thin cuts of rock are mounted on glass slides and studied under microscopes.

Study of the forams, as well as of the brachiopods and corals has shown the Redwall Limestone to have been deposited during the Early and Middle Mississippian Period. The fossils and the nature of the rock itself suggest that the Redwall Limestone was deposited in a shallow tropical sea of great lateral dimensions.

Just as thick sequences of rock are divided into "formations" composed of similar rock types, formations may be subdivided into still-smaller units called "members" if it is practical and useful to do so. The Redwall Limestone has been separated into four members for ease of studying the more-detailed geology of the formation; each of the members is named in the same fashion as formations. The lower member is a massive cliff of limestone or dolomite overlain by the second member, which contains abundant beds and nodules of chert. The top of the second member is marked by an eroded surface, or unconformity, of small time significance, called a "diastem." This particular diastem is interesting because it shows that while the Redwall was being deposited in a shallow tropical sea, a brief lowering of sea level occurred, permitting the erosion of the top of the second member, which was then at the upper surface of the sequence. The brief cessation of sedimentation is marked in the Grand Canyon by small channels eroded into the second member, and later buried by the deposition of lime sediments of the third and fourth members. As we will see later, this diastem can be traced over hundreds of miles of the Colorado Plateau by a thin layer of pebbles at about this same position in the formation.

Surprise at Surprise Canyon

It would appear that any rocks deposited in the later half of the Mississippian Period were eroded from the Grand Canyon region, and that rocks of Pennsylvanian age overlie the Redwall Limestone. During geologic mapping by George Billingsley in remote parts of western Grand Canyon in 1969, however, deep erosional channels filled with a variety of sedimentary debris were discovered at the tope of the Redwall Limestone. Because the exposures occur in remote tributary canyons to the Colorado River, the channel- and valley-fill deposits are the least known of the Grand Canyon. The newly discovered deposits were named the Surprise Canyon Formation for exposures in Surprise Canyon, and appropriate surprise, by Billingsley and Stan Beus in 1985.

The newly named formation consists of basal conglomerate beds in some outcrops, overlain by sandstone, shale and limestone beds that contain abundant marine fossils. Exposures are variably distributed throughout Grand Canyon in what appear to be a dendritic erosional pattern that extends from east to west into the Cordilleran seaway. The distribution of the channel system, and the contained fossils, indicate that the formation was deposited in estuarine channels in latest Mississippian time (Chesterian).

A Brief Intermission (Pennsylvanian Period)

Deposition of the Redwall Limestone ceased in about mid-Mississippian time, and no geologic record of the remainder of Late Mississippian time is known to occur in Grand Canyon. Pennsylvanian and Early Permian sedimentary rocks directly overlie the Redwall in the Grand Canyon, superimposing red sedimentary rocks onto the massive limestone. A deeply weathered surface separates the Mississippian rocks from overlying beds throughout

The strata of Permian and Pennsylvanian age in Grand Canyon as seen from the Kaibab Trail. The top of the lower cliff is the base of the Supai Group, which forms the ledgy slopes in the middle of the photograph. The prominent slope-forming unit about one third of the way down from the top is the red Hermit Shale, and the large vertical cliff above it is the result of windblown deposits of the Coconino Sandstone. The uppermost slope is carved in the Toroweap Formation, and the top cliff represents the approximate position of the Kaibab Formation.

the Colorado Plateau and much of the Rocky Mountain region, and a similar surface of weathering and soil formation marks the boundary in Grand Canyon. This erosional unconformity is represented by deeply weathered limestones, sinkholes, and caverns in the upper part of the formation caused by the removal of limestone by ground waters that dissolved large parts of the relatively soluble rock in Late Mississippian and Early Pennsylvanian time. These caverns and sinkholes are now filled with red sediments and soils from the upper surface or overlying beds, indicating that the weathering occurred before Permian time rather than being modern. The presence of this unconformity over such a broad area suggests that the sea withdrew from the entire Colorado Plateau and Rocky Mountains region in Late Mississippian time, and a broad weathering surface formed in a moist, subtropical climate, producing a lateritic soil over much of that region. (In lateritic soil, iron and aluminum remain in the surface layer but silica is dissolved and removed.)

It is uncertain whether sediments of Pennsylvanian age were deposited throughout the entire Grand Canyon region, although fossils of marine-dwelling organisms (brachiopods) of Early and Late Pennsylvanian age have been found in the western parts of the canyon. Similar limestones to those containing the brachiopods are present between red shales across almost the entire length of the canyon, but they are unfossiliferous in the eastern occurrences and therefore cannot be dated with any accuracy. These limestones in the basal Supai Group are of Pennsylvanian age, and they represent only thin remnants of the much thicker Pennsylvanian marine deposits that literally surround the Kaibab uplift (Grand Canyon region). Rocks of Middle Pennsylvanian age (Desmoinesian) have not been identified in the Grand Canyon. Strata that contain fossils from the entire Pennsylvanian Period are one thousand to several thousand feet thick toward the west in the Cordilleran seaway and toward the northeast in the Four Corners region, and only slightly thinner deposits surround the Kaibab uplift. The limestones in question in eastern Grand Canyon are no more than a few feet thick and represent only small portions of Pennsylvanian time. This situation again suggests that the Kaibab uplift was slightly positive during Pennsylvanian time relative to surrounding regions, recording early development of the broad arch at this time.

Red Beds Inherit the Earth (Permian Period)

The upper half of the layered rocks in Grand Canyon is composed of gentle slopes of red shale and siltstone, punctuated by light-colored cliffs of sandstone and limestone, all of Permian age. This colorful sequence of sedimentary rocks provides much of the artistic appeal of the canyon, and the uppermost layers constitute the cap rock that rims the canyon and supports the flat Kaibab and Coconino plateaus on either side of the canyon. It is an interesting series of rock layers, because the red sediments were probably deposited in continental environments such as floodplains along rivers, deltas, lakes, and dune fields. Similar deposits of Permian and younger age are very extensive on the Colorado Plateau, providing much of the scenic beauty of the entire region.

The lower strata of this picturesque series of rocks make up the Supai Group. The apparently continuous sequence was recently subdivided into four formations that are separated by minor unconformities. The lower unit named the Watahomigi Formation usually contains thin limestones interbedded with red shales and siltstones. The next higher formation, named the Manakacha, is separated from the overlying Wescogame Formation by a diastem; both are mainly redbed sequences. All three of these formations grade westward into marine strata of Pennsylvanian age. An erosional unconformity marked by a thin layer of conglomerate at the top of the Wescogame is thought to be the temporal boundary between the Pennsylvanian and Permian periods. More thin redbeds overlain by a thick sequence of sandstone has been called the Esplanade Formation (Permian age) as the resistant sandstone forms a prominent topographic bench known as the Esplanade throughout the length of Grand Canyon. These four formations of similar sedimentary rocks make up the Supai Group. The ledgy slopes of the lower three formations are primarily redbeds whose sediments were deposited on broad flood plains of rivers in an arid climate. The headwaters and source of the sediments was in high mountains of the Ancestral Rockies range in Colorado, where alpine peaks were being torn down by erosion during this time. The red coloration is due to a concentration of highly oxidized iron known as hematite, which makes up much of the finer sediments and coats the coarser sand grains. The pigment may have been deposited as red mud formed by the weathering of the granites in the mountains, or it may have altered to red from drab-colored muds by the intense oxidation of iron-bearing minerals after deposition along the flood plains. In either case, the muds were exposed to strongly oxidizing conditions after deposition to produce or retain the red hematite.

Toward the west and northwest, the upper red beds interfinger with limestones containing fossils of marine organisms of Lower Permian age.

These rocks of marine origin are known as the Pakoon Formation in the western extremities of Grand Canyon, and the Elephant Canyon Formation in central Utah. The upper sandstone cliffs of the Supai Group, named the Esplanade Sandstone by early geologists in the region, are cross-bedded (many depositional surfaces are inclined to the horizontal), indicating that they were deposited by aqueous currents as sand waves or small sand bars; many geologists would argue that these are windblown dune deposits. The Supai Group was probably deposited in a rather arid climate, as shown by the nature of the plant fossils (mainly primitive ferns) found in it. Footprints of primitive four-footed, shortlegged amphibian animals have been found in the red shales. They measure several inches in length, with three- and five-toed varieties being rather common. These clumsy salamander-like animals plodded along the river floodplains in search of food; their bones have been discovered in the Monument Valley region to the east.

Gullies and channels were eroded into the top of the Supai sandstone ledge in Grand Canyon by ancient streams, and these were infilled and eventually covered by more red muds of the overlying Hermit Shale. These muds, like those of the Supai, were deposited on floodplains of sluggish Permian rivers, as shown by the channels, ripple marks, raindrop impressions, and fossil leaves and tracks of amphibians found in this formation. Fossils of about thirty-five species of leaves have been described from the Hermit Shale, indicating a semiarid climate during deposition of the formation. The plants are primarily ferns and cone-bearing plants of diminutive proportions, with a noticeable paucity of moist-climate varieties. A few insect wings occur with the fossil leaves. The Hermit Shale is the highest of the red beds preserved in Grand Canyon and forms a strong topographic notch.

The top of the Hermit Shale is marked by a sharp, smooth contact with the overlying light-colored deposits of the Coconino Sandstone. The massive cliffs of this formation are seen as a prominent white band near the top of the canyon walls. The bedding surfaces within this formation are not horizontal, as in most of the strata, but are instead inclined at very steep angles in a most erratic manner. Upon close examination, it is seen that the sand grains of the Coconino are very white and quite well rounded, and that the surfaces of the individual grains are pitted and frosted. The grains are rather fine in size as sand goes, and are all about the same size. The only traces of life in the formation are tracks and trails of worms and salamander- or lizard-like creatures such as may commonly be seen to cross smooth bedding surfaces. These features are quite characteristic of modern wind-deposited dunes.

It is believed that the Coconino Sandstone is an ancient example of a desert or coastal field of sand dunes that were frozen in place by the dampening of ground waters. The steeply inclined and erratic bedding surfaces represent the sites of sedimentation on the steep leeward slopes of the dunes, and the high degree of sorting, rounding, and frosting of the grains was due to the normal processes of wind transportation. As the individual dunes were swept across the coastal desert by the winds, the internal cross-bedding characteristics were formed by eroding sand from the windward side and redepositing it in succeeding layers on the steep leeward side. Studies of these lee-slope deposits suggest that the prevailing winds were from the north during deposition of the Coconino. South of Grand Canyon, however, the sand grains become gradually coarser in size and more angular in outline, strongly suggesting that the sand was not transported so far in that direction and probably came from the south of Grand Canyon in central Arizona. If the winds that deposited the Coconino dunes blew out of the north, and the sand came from the south, the sand must have been first distributed by streams northward from the source area and later blown up from the stream beds into sand dunes.

The sea returned twice in Middle Permian time to deposit marine limestones and shales of the Toroweap and Kaibab formations, which cap the Grand Canyon section. These formations are seen as the uppermost ledges and cliffs immediately beneath the rim, but above the white cliffs of the Coconino. The surface upon which the lower Toroweap sediments were deposited as the Middle Permian sea advanced was a very smooth plain on the top of the Coconino Sandstone. Although the contact has the appearance of an unconformity because of the sharp and abrupt change of rock type, it is more likely that the surface was produced by the reworking of the top of the Coconino by the advancing sea.

The Toroweap Formation is divided into three members at most localities within Grand Canyon. The lower member is a thin red and yellow sandstone that was probably deposited on the shoreline from sands of the Coconino over which the sea was advancing. This unit grades upward into a middle member composed of massive limestone beds that thin eastward through the canyon from about two hundred feet in the west to zero just east of the canyon along the Little Colorado River. This member contains numerous fossil brachiopods of a particularly fat and thick-shelled variety. It was deposited when the Toroweap sea was at its maximum, easternmost development. As the sea slowly retreated from the Grand Canyon region, a series of red shales and siltstones were deposited above the middle limestones, forming a strong notch where it is weathered out in the canyon walls below the cap rock. These red beds change eastward into windblown sand deposits much like the Coconino in the vicinity of the Little Colorado River and Walnut Canyon National Monument. Thus, all evidence points to the Toroweap sea having advanced from the west toward a shoreline just east of the present-day Grand Canyon, and then having backed out briefly into the Cordilleran region toward the west. The upper red-bed member was attacked by a brief period of erosion that followed the retreat of the sea, producing a gentle unconformity at the top of the formation.

The sea again advanced over the Grand Canyon region in Middle Permian time, depositing the limestones that cap the rim of the canyon. This unit, the Kaibab Formation, is mainly a limestone that was deposited in warm, shallow seas. In Grand Canyon, the Kaibab limestones contain abundant marine fossils, which include brachiopods, clams, crinoids, and horn corals. Eastward the Kaibab becomes more sandy until it is a brown sandstone deposited along a north-south shoreline at the eastern margin of the canyon. This sandy trend or "facies" contains a brackish-water clam and snail fossil fauna. An upper member of the Kaibab Formation is a fossiliferous marine limestone and red-bed unit that was deposited as the Permian sea retreated for the last time toward the Cordilleran seaway, which lay to the west. This upper unit contains abundant fossils of Permian trilobites, clams, snails, and cephalopods (coiled and chambered marine mollusks), with a few brachiopods in some outcrops. It also contains beds of gypsum in the region west of the canyon, which were deposited in shallow evaporating ponds as the sea retreated.

No deposits of Late Permian age are known to occur in the Grand Canyon country.

Post-Paleozoic History of the Grand Canyon

Although the Kaibab Formation is the uppermost rock unit that is evident along Grand Canyon, it was not the last event in the history of the region. At least ten thousand feet of younger rocks were deposited above the Kaibab in Mesozoic time, as shown by the remnants of these rocks that have been preserved surrounding the Kaibab uplift (Grand Canyon plateau region). That these rocks were once extensive over the Grand Canyon region is shown by remnants of red sandstones and shales of Triassic age preserved

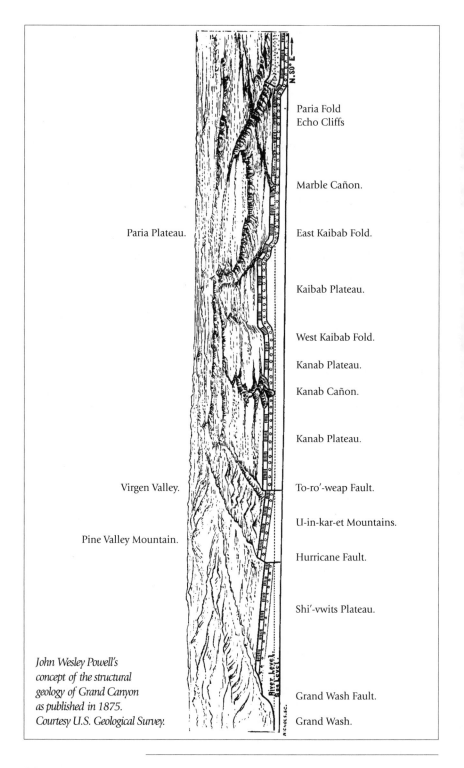

Paria Fold
Echo Cliffs

Marble Cañon.

Paria Plateau.

East Kaibab Fold.

Kaibab Plateau.

West Kaibab Fold.

Kanab Plateau.

Kanab Cañon.

Kanab Plateau.

Virgen Valley.

To-ro'-weap Fault.

U-in-kar-et Mountains.

Pine Valley Mountain.

Hurricane Fault.

Shi'-vwits Plateau.

John Wesley Powell's
concept of the structural
geology of Grand Canyon
as published in 1875.
Courtesy U.S. Geological Survey.

Grand Wash Fault.

Grand Wash.

The eastern limits of Grand Canyon as seen from the air. The strata in the middle distance and beyond can be seen to bend abruptly down toward the right, or east, to form the East Kaibab monocline, which bounds the Kaibab uplift on the east. Toward the upper right of the view is Marble Canyon, which leads into Grand Canyon in the center of the photograph; the canyon in the lower right is the lower termination of the Little Colorado River canyon.

above the Kaibab in two localities. At Red Butte, a low rounded hill some fifteen miles south of Grand Canyon Village, the red beds were preserved from later erosion by a hard cap of lava. The other erosional remnant of younger strata is at Cedar Mountain, a mesa that can be seen east of Desert View, at the east end of the canyon. The nature of this great thickness of post-Permian sedimentary rocks will be discussed in a later chapter, but it is important at this point to realize that such a column of strata once existed.

We have seen that the Grand Canyon region was somewhat uplifted at various times during Paleozoic sedimentation. At some later time, which cannot be fixed with much accuracy but was probably at about the end of the Mesozoic, the entire Grand Canyon region was uplifted into a great, gentle fold. The eastern limb of this arching of the strata can be seen just east of Grand Canyon, where the strata bend over sharply and dip steeply into the subsurface toward the east along the "Kaibab monocline." The western flank of the broad fold is much more gentle and barely discernible, for the strata dip very gently westward throughout the length of Grand Canyon. This broad upfolded structure is known as the Kaibab uplift.

At the time the principal arching was taking place, the very thick post-Permian strata were already deposited and were upfolded along with the older rocks. After the folding of the Kaibab uplift had commenced, erosion attacked the higher areas that were raised by the folding, for erosion always works first and hardest against the higher regions, to level the surface of the earth if at all possible. The end result was that the younger strata were eroded away from the entire top of the Kaibab uplift down to the level of the Kaibab Formation. The reason that the Mesozoic rocks were removed and not the Paleozoic is that the younger rocks were mainly soft sandstones and shales, which were easily removed by normal weathering and transportation processes, while the Kaibab limestones were very hard strata, highly resistant to erosion. So the softer, less-resistant rocks were removed from the top of the Paleozoic section.

The main erosive forces of the Colorado River did not reach the top of the harder Paleozoic strata until about late Tertiary time, only a few million years ago. As it attacked the hard layers it carved a steep-walled canyon into the uplifted plateau, producing a canyon that cuts directly across the high plateau. The river did not actually attack the plateau from the east and tear a hole into it, but, instead, the canyon was gradually cut downward into the top of the upfolded layers as the river lowered its bed into the plateau by normal erosion. Although it is hard to believe that the Colorado River could have cut such a chasm, a trip down to the river in the Grand Canyon will convince even the most skeptical that it is a mighty, roaring torrent that can do a great deal of destruction. Even at that, if it were not for the extra sand-papering effect of the sand and gravel carried along by the river, it could not do the efficient downcutting job that produced the canyon. Also, it must be realized that for at least a million years this roaring torrent has been actively tearing at the floor of the canyon at approximately the same rate as today. If the river removed a few grains from the river bottom each day, it would add up to an astonishing amount of erosion in a million years' time. While the

Eastern Grand Canyon as seen from the air. Notice the very broad profile of the canyon formed by canyon-widening processes.

river is solely responsible for the downcutting of the channel, the canyon is being actively widened by the erosion effects of wind, rain, freezing and thawing, and the removal of the material toward the river by the forces of gravity. These processes have produced a canyon that is a mile deep and an average of ten miles wide.

Where has all the eroded material gone? Before construction of Glen Canyon Dam upstream from Grand Canyon in the early 1960s, the river transported about a million tons of sand and silt past any particular point every day. This million tons of sediment was originally transported to the Gulf of California, where it was deposited as marine and tidal-flat deposits to become sedimentary rocks to challenge the geologists of some future time. When Boulder Dam was constructed downstream from Grand Canyon, these sediments were deposited in Lake Mead, and began filling the reservoir at an alarming rate. Since Glen Canyon Dam was built, largely to slow the filling of Lake Mead, sediments are accumulating rapidly in Lake Powell above the dam, and the Colorado River runs green and clear much of the year through Grand Canyon. This situation can only end in the development of vast corn fields for the Navajo people of the next geologic period.

CHAPTER TWO

JOURNEY INTO THE GEOLOGIC DARK AGES

THE GRAND CANYON BY BOAT

Early pioneers traveling in the vicinity of northern Arizona encountered an almost insurmountable natural obstacle if they were going to or from southern Utah. The canyons of the Colorado River, although spectacular and awe-inspiring, were truly a headache to the early settlers, who came into the region mainly from the Mormon settlements in the north. Only a few crossings were known, and these were widely spaced geographically and extremely hazardous to negotiate with horses and wagons. One of the major crossings was at Pierce Ferry, near the present site of Lake Mead and at the west end of Grand Canyon. Another relatively easy crossing was at Hite, Utah, which is just below Cataract Canyon, a distance of some three hundred miles upstream, near the mouth of the Dirty Devil River. In between lay only vertical canyon walls with sheer drops of several hundred feet guarding the stubborn river. Then, in the 1850s, Jacob Hamblin rediscovered a crossing that had been used by Father Silvestre Velez de Escalante and his party in 1776 near the mouth of the Paria River at what is now the transition between Glen and Marble canyons. This proved to be a usable crossing, and John D. Lee built a ferry to facilitate its use. Lee's Ferry became the focal point for all river crossings between the widely spaced Pierce and Hite ferries, largely as a consequence of the geologic setting of the unique location.

Glen Canyon extends from Lees Ferry upstream for a distance of about 150 miles, carved into the relatively flat-lying layers of massive red sandstones of Jurassic age. The Mormon settlers attempted to establish a crossing within this vertically walled trench at Hole in the Rock, but it was such a treacherous matter to lower and raise the teams and wagons over the cliffs that it was soon abandoned. The site of Glen Canyon Dam, near Page, Arizona, now provides a crossing of the spectacular canyon that would have been the envy of the early pioneers; however, this is still the only modern crossing between Hite and Navajo Bridge, immediately downstream from Lees Ferry.

The nearly horizontal Triassic and Jurassic strata bend upward abruptly toward the west at Lees Ferry and along the north-south trending Echo Cliffs, and are carved away from the tops of the uplifted plateaus that Marble and Grand canyons truncate. Where the Jurassic sandstone cliffs are uplifted along the Echo Cliffs flexure, an older Triassic series of soft-weathering red shales known as the Moenkopi and Chinle formations rise to the surface to form broad, gentle valleys adjacent to the Colorado River. It is at this widening of the canyon by virtue of the exposure of the shale formations along a major fold in the earth's crust that Lees Ferry was founded.

The canyon of the Colorado River again becomes virtually impenetrable immediately below Lees Ferry, where the much more resistant limestones of the Permian Kaibab and Toroweap formations rise to the surface along the fold and form vertical-walled bastions to Marble Canyon. It is this very narrow gorge that Navajo Bridge now spans, forming the only other crossing for vehicles between Boulder Dam (near Las Vegas, Nevada) and Hite. A second bridge, widened and improved, adjacent to the historic Navajo Bridge built in 1928, was completed and opened for traffic in 1995.

The relatively easy approaches to Lees Ferry used by the pioneer wagons can still be seen from the boat landing, where modern adventurers rig their rubber rafts and pontoons for the two-hundred-mile journey through Marble and Grand canyons. Each trip is anticipated with the spirit of adventure of a major exploration expedition by its participants. But modern river runners cannot experience anxiety like that of Major John Wesley Powell and the motley crew who passed this point on August 5, 1869, in three small wooden boats on the first penetration of the mysterious canyons. To these explorers the canyons contained unknown dangers and treacheries, the possibilities of countless major rapids and even impassable waterfalls, and worse, perhaps starvation, if their meager rations should be lost in a mishap in the river's torrents or the canyon could not be completely traversed. Thanks to the determination and courage of these men and others such as those on Powell's second expedition in 1872, and the later Stanton, Kolb brothers, Stone and Galloway expeditions, the river and its treacheries are now well explored and well understood.

Because of the relatively new Glen Canyon Dam, even the amounts of water to be challenged may be predicted beforehand, for the stage of the river, once so unpredictable and critical to safe passage of the rapids, is now determined by the amount of water released by the dam authorities. Even with these modern advantages, a boat trip through Marble and Grand canyons is still one of the greatest remaining adventures and one of the most inspiring experiences to be had in the western world.

We push off from the boat ramp at Lees Ferry in a fleet of rubber pontoons with great expectations for the journey ahead. There are several professional river outfitters who are approved by the National Park Service to conduct trips through Grand Canyon, and all are safe and reliable. We are traveling with Ted Hatch, whose father, Bus Hatch, pioneered the use of rubber boats in white-water travel. It is possible to run Grand Canyon without professional boatmen, but the requirements of the Park Service are properly so strict that it is not practical for most people to attempt it. Our pontoons are loaded with camping equipment and culinary delights to be savored along the way. Our cameras are loaded and we are ready for the task ahead, when the magnificent canyon spectacles will require sudden use of the camera for long periods. Our boatmen are primed for the challenge of the river.

The limestone cliffs of the Permian formations rise rapidly onto the elevated plateau as we leave Lees Ferry behind. Because the river is cutting downward into uplifted strata of increasing elevation, we will be traveling through progressively older and older layers as we follow the course of the river. Instead of seeing the geologic formations in the order in which they were deposited and in which they were described in the previous chapter, we will be seeing them in the sequence in which they would be penetrated by a drill: From the top down. Within the first couple of miles, the Colorado River has cut a very narrow canyon into the Kaibab-Toroweap formations, and the windblown deposits of the Coconino Sandstone come into view at river level. Because of the vertical walls of the now-shallow gorge, the three Permian-age formations are not very distinctive, for they do not separate into the classic stairstep topography as in Grand Canyon proper. Furthermore, the formations are not nearly so thick here, and actually pinch out just east of Lees Ferry. The canyon becomes continuously deeper, and in the next couple of miles the cliffs reach formidable proportions as we pass well beneath Navajo Bridge and the highway. Then, as the flexure in the rocks is completed, the strata level off and the deepening of the canyon and the intrusion of the older strata slow appreciably.

The first major rapids of the journey are encountered at mile 8, where Badger Creek enters the gorge. As in most of the rapids along the Colorado River, these form where a tributary stream carries loads of gravel and boulders into the main canyon in excess of the amount the Colorado River can remove. The consequent overloading of the channel with coarse debris during times of flash floods constricts the channel, and rapids form.

Red beds of the Hermit Shale rise to river level at this point and can be traced downstream to the next major rapids at Soap Creek, where the top of the underlying formations of the Supai Group soon becomes obvious. The canyon widens somewhat as the soft Hermit strata form the lower slopes beneath the more resistant Coconino, Toroweap, and Kaibab formations. Boulders of the more resistant beds from above fall onto and partially mask the Hermit slopes as the canyon begins to widen itself, and the difference between the deepening of the canyon by the downcutting of the river and the widening by gravity slumping becomes evident.

Below Soap Creek rapids, the canyon again narrows to a rugged defile as the river attacks the resistant layers of the upper Esplanade Sandstone of the Supai Group. The Esplanade is here very thick and highly cross-stratified, more like the massive Cedar Mesa Sandstone of the Canyonlands country than the Supai of Grand Canyon. The Supai Group is exposed in the canyon walls for about twelve miles, where the details of its character can be carefully examined amidst scenic views of the overlying Permian layers of sandstone, limestone, and red beds.

The colorful Permian strata are left behind at about Tanner Rapids (mile 24.5), although they will follow our progress from their elevated perch throughout the remainder of our trip. Here the Redwall Limestone is encountered at river level, and the unconformity at its upper surface can be readily studied. The canyon again narrows markedly as the river carves its path into the massive, hard layers of the deposits formed in the Mississippian sea that once covered this arid stretch of the west. That the Redwall was deposited in an ancient sea is made obvious by the many fossils of marine animals that are exposed in the limestone cliffs.

The sheer inner gorge deepens gradually, and the limestone walls rise majestically as the river carves its way into the Redwall Limestone. The soft light of evening plays across the vertical cliffs, displaying delicate shades of pink and red, giving an eerie feeling to the ever-deepening, ever-changing walls of this magnificent canyon. At many places the river occupies the entire canyon floor and the view upward is like that from the bottom of a well.

Powell wrote: "August 9, 1869. And now, the scenery is on a grand scale. The walls of the cañon, 2,500 feet high, are of marble, of many beautiful colors, and often polished below by the waves, or far up the sides, where showers have washed the sands over the cliffs." He goes on: "We have cut

Air view, looking northeast, of upper Marble Canyon, cut into the nearly flat plateau, and the Vermilion Cliffs in the left background. The section of the canyon exposed at the bottom of the photograph consists of an upper cliff of the combined Kaibab, Toroweap, and Coconino formations in descending order, a broad slope of the Hermit Shale, and a lower cliff and ledgy slope made up of the Supai Group.

through the sandstones and limestones met in the upper part of the cañon, and through one great bed of marble [the Redwall Limestone] a thousand feet in thickness. In this, great numbers of caves are hollowed out, and carvings are seen, which suggest architectural forms, though on a scale so grand that architectural terms belittle them. As this great bed forms a distinctive feature of the cañon, we call it Marble Cañon."

Nearing the base of the formation, Ted Hatch leads us into a narrow tributary canyon and proudly points to several straight shells of a long-extinct variety of cephalopod that have been partially eroded from the limestone by the tributary torrents during innumerable past flash floods. The giant mollusks that inhabited these shells, measuring over a foot in length, must have been the masters of the sea floor here in Lower Mississippian times.

"Riding down a short distance, a beautiful view is presented," wrote Powell. "The river turns sharply to the east, and seems enclosed by a wall, set with a million brilliant gems. What can it mean? Every eye is engaged, everyone wonders. On coming nearer, we find fountains bursting from the rock, high overhead, and the spray in the sunshine forms the gems which bedeck the wall. The rocks below the fountain are covered with mosses, and ferns, and many beautiful flowering plants. We name it Vasey's Paradise, in honor of the botanist who traveled with us last year."

We pull into shore just above Vasey's Paradise at the mouth of Paradise Canyon and make camp for the night beneath the now brilliantly illuminated pink-, red-, and orange-painted walls of the Redwall Limestone. The view both up and down the canyon at this point is one of intense beauty. One gets the feeling of being swallowed up by an ever-deepening abyss in the earth's crust from which there is no escape. It is here, also, that one begins to get the strange sensation that our journey is not merely a trip down the river, but rather a trip back into the dark, past ages of geologic time.

We have already traveled back in time some 350 million years, as if in a large silver time machine of the future in the form of a rubber pontoon, with an unknown interval of time still to explore, by frozen mud and oolite sand, with their contained remains of ancient sea life. Tomorrow, perhaps, we will continue our journey through the mystery and unknown events of the great Devonian-Silurian-Ordovician hiatus and beyond into other seascapes of 500 to 570 million years ago, in Cambrian time. We sleep under the magic of a star-filled sky that would tax the imagination of a poet.

In the morning, after exploring the Redwall Limestone for fossils and Indian caves, one of which contains an Indian skeleton, we push off again drifting past the crying walls of Vasey's Paradise and the magnificently undercut retreat of Redwall Cavern, toward the ravished canyon walls of the Marble Canyon Damsite (mile 40). By the time we arrive at the proposed site of the dam that would have buried this canyon wonderland beneath a couple hundred feet of stagnant water, we have crossed the unconformities that separate the Redwall Limestone from the underlying Devonian and Cambrian strata.

Several well-developed and well-exposed channels cut into the top of the Cambrian Muav Limestone and filled with Devonian dolomite of the Temple Butte Formation can be seen in the vicinity of the damsite on either side of the canyon. These remnants of an otherwise missing chapter in the history of Grand and Marble canyons are exposed at or near river level, where they can be readily studied and the significant erosional unconformities can be fully appreciated. At localities where the channels are present, the Mississippian strata overlie beds of Devonian age, with only very little of Devonian time missing at the intervening unconformity. At nearby localities where no Devonian channels occur, the Redwall rests on dolomites of Cambrian age—an unconformity representing all of Devonian, Silurian, and Ordovician time.

It is at about this important geologic horizon that the Bureau of Reclamation has raped the canyon walls with its probes and test holes and hideous steel scaffoldings, left in place, until recently, as a reminder of what could have happened to the canyon and what still may happen at some future time when we are caught asleep by the dam builders. [Although still apparent, scaffolding has been removed by the Park Service, and tunnels have been closed.]

We make our second camp near President Harding's Rapid at the top of the greenish-colored shaly beds of the Bright Angel Shale of Late Cambrian age. The illusion deepens that we are traveling back in time, for we have lost all contact with the everyday world and continue to find older and still older relics of ancient seas. In the Muav Limestone today we have discovered trilobites, the segmented crablike animals that dominated the Cambrian seas some 500 million years ago. What clues to the even more ancient environments and their inhabitants will tomorrow bring?

After drifting down the Colorado River for about eight miles on the morning of the third day, we come to Nankoweap Rapid and make a prolonged stop to examine Indian ruins and fossil trilobite and brachiopod remains in the Bright Angel Shale. It is at Nankoweap Canyon that the Colorado swings toward the south and flows along a second great flexure in the earth's crust known as the East Kaibab monocline, which elevates the strata another two thousand feet toward the west. This uplifted fold is the large Kaibab uplift, which is dissected by Grand Canyon proper. The uplift is capped by the Kaibab Limestone, all younger strata that may have originally been deposited in the area having been stripped off the elevated plateau by relatively recent erosion.

The canyon here takes on an asymmetrical shape, since the river has tried to slip down the dip of the flexure toward the east and consequently has carved a very steep eastern wall, leaving a much less abrupt west wall. These relationships are far more obvious when seen on topographic and geologic maps than from the deeply entrenched position of the canyon floor,

Aerial view of the canyon of the Little Colorado River just above its confluence with Grand Canyon. The break in slope in the middle distance is a small monocline known as the Grandview monocline, where the Paleozoic strata climb higher toward the west and southwest onto the Kaibab uplift. The plateau is underlain by the Kaibab Formation, with the canyon being incised as deep as the basal Supai Group.

where the immediate canyon assumes the role of the entire universe. The river marks the eastern boundary of the original Grand Canyon National Park from Nankoweap Canyon to the mouth of the Little Colorado River about ten miles farther south.

The Colorado River, once deep red and seemingly thick from the heavy load of suspended matter that it carried down from the Rockies, now runs as clear as a mountain brook through Marble Canyon. Glen Canyon Dam has created an extensive settling basin and sediment trap, which, in time, will become a magnificent red canyon filled with mud. Grand Canyon Park rangers used to say that the river in Grand Canyon was "too thick to drink and too thin to plow." The river now is thin enough to drink in Grand Canyon and it will soon be thick enough to plow in Lake Powell above Glen Canyon Dam.

The situation changes abruptly when the mouth of the Little Colorado River is reached during the time of spring runoff. The "Colorado Chiquito," as Powell called it, runs muddy and red in springtime, forming an abrupt wall of turbid, red water as it encounters the clear Colorado. Below this milestone in the river trip the water will remain a deep reddish-brown color until it reaches the next reservoir, at Lake Mead below Grand Canyon. Although the Little Colorado is a main tributary to the Colorado, there is no rapid at the confluence. The reason is that the sediments carried by the Little Colorado are restricted to fine sands and muds because of the low gradient of the lesser stream, and the boulders that form exhilarating rapids cannot be transported to the junction. The sand and mud bars formed at the mouth of the smaller stream are easily removed by the Colorado, and any potential obstruction to the flow of the main river is constantly being scuttled.

Also, in the vicinity of the confluence of the Little and big Colorado rivers, the basal unit of the Paleozoic rock column is beautifully exposed. The Tapeats Sandstone forms cliffs on either side of the canyon that are somewhat peculiar in that they are composed of innumerable thin ledges of sandstone, each cross-stratified on a small scale. It is not difficult to see that this type of sedimentary structure is not formed in rivers or windblown dunes, but must instead be the result of marine processes along or very near a shoreline. With a little imagination, one can visualize sandy beaches of more than 500 million years ago, and, because of the paucity of fossil forms in the deposits, one can appreciate how destitute a clam digger or a pearl farmer would have been in Middle Cambrian times.

The very ancient and the modern are tragically brought into juxtaposition in Chuar Butte, just across the Colorado River to the west from the mouth of the Little Colorado. There the wreckage of two giant airliners that collided in mid-air over Grand Canyon came to rest among the ledges and crags, strewing metallic debris and human life over the canyon walls. The area is now set aside as a memorial to those dozens of people who lost their lives in the modern tragedy.

Vishnu's Domain

Upon leaving the mouth of the Little Colorado River, the course of the canyon wanders to the west and crosses into the Kaibab uplift and into Grand Canyon proper. One can appreciate the anxieties of Major Powell and his men as they entered the canyon on the initial probe by reading Powell's diary: "August 13, 1869. We are now ready to start on our way down the Great Unknown. Our boats, tied to a common stake, are chafing each other, as they are tossed by the fretful river. They ride high and buoyant, for their loads are lighter than we could desire. We have but a month's rations remaining.... The lighting of the boats has this advantage: they will ride the waves better, and we shall have but little to carry when we make a portage.

"We are three-quarters of a mile in the depths of the earth, and the great river shrinks into insignificance, as it dashes its angry waves against the walls and cliffs, that rise to the world above; they are but puny ripples, and we but pigmies, running up and down the sands, or lost among the boulders.

"We have an unknown distance yet to run; an unknown river yet to explore. What falls there are, we know not; what rocks beset the channel, we know not; what walls rise over the river, we know not. Ah, well! we may conjecture many things. The men talk as cheerfully as ever; jests are bandied about freely this morning; but to me the cheer is somber and the jests are ghastly.

"With some eagerness, and some anxiety, and some misgiving, we enter the cañon below...."

The river enters the narrow defile carved through the Tapeats Sandstone, but within a couple of miles the canyon opens up into the broad valley of upper Grand Canyon as the relatively nonresistant younger Precambrian red beds of the Grand Canyon Supergroup rise into view and are exposed to the processes of erosion. The contact between the Paleozoic rocks and the underlying Precambrian sedimentary strata is one of angular unconformity. The older rocks were subjected to the earlier stresses within the earth's crust, which gently folded and faulted them, but the processes were not severe enough to alter the rock type to metamorphic rocks.

Indeed, these ancient strata, which were deposited perhaps a billion years ago, look every bit as fresh as the red Triassic rocks at Lees Ferry, and could be confused with those rocks if their position beneath the hundreds of feet of strata of Paleozoic age were not understood. The red and brown siltstone, sandstone, and shale beds of the Grand Canyon Supergroup, which attain a thickness of some 13,000 feet, were first exposed to erosion prior to Cambrian time. The gentle folds and faults and the resulting inclined bedding were partially eroded to a gently rolling surface in this area when the marine waters of the Cambrian Tapeats sea began its transgression. The basal Paleozoic sands were deposited in a more or less horizontal attitude over the inclined older beds, forming the angular discordance at the contact that is so obvious along the river at this point.

Faults and folds are numerous in the upper Precambrian red beds, but they do not affect the Paleozoic strata, telling us that the folding occurred prior to the deposition of the Cambrian sand. We camp in the peaceful, broad canyon among the vermilion slopes and escarpments, and admire the remarkable preservation of ripple marks, cross-stratification, and fossil mud cracks in these ancient sediments known as the Dox Formation. The red beds were deposited on floodplains and in river channels at least a billion years before the appearance of man on earth.

The entire sequence of Paleozoic formations can be distinguished from the open, rolling valley; the low, flaggy cliffs of the Tapeats Sandstone at the first unconformity; the greenish, gentle slopes of the Bright Angel Shale capped by the vertical escarpment of the combined Muav through Redwall limestones; the red ledgy slopes of the lower Supai Group grading upward into cliffs of the Esplanade Sandstone; the red bench formed on the outcrop of the Hermit Shale; and the overlying cliffs and alternating slopes of the Coconino Sandstone, the Toroweap Formation, and the rimrock composed of the Kaibab Limestone. The tower of Desert Viewpoint, on the south canyon rim, stands out boldly against the desert sky.

On the fourth morning we drift downstream through the rolling red valleys, past fault blocks accentuated by the presence of basalt sills interspersed between the ancient red beds, and past Unkar Rapid. Indian ruins dot the gentle valley slopes and terraces, and the mood of the canyon is one of peace and serenity. The strata are dipping gently toward the east now, and we proceed into older rocks as we continue downstream. The canyon bares it teeth once more as a thick, very hard sandstone bed within the Grand Canyon Supergroup, known as the Shinumo Quartzite, rises to the surface and forms a foreboding abyss in our path. The threatened dangers do not materialize, however, until the river passes entirely through the quartzite layer and emerges into the widening canyon. There, still more red strata of the Grand Canyon Supergroup, the Hakatai Shale, appear at its base.

At this point, at the head of the treacherous Hance Rapid, we have traveled about seventy-five miles from our starting point at Lees Ferry, downward through millennia of geologic time until we find ourselves at the bottom of this ever changing gouge in the earth nearly a mile deep. We are reminded of this proximity to the hot core of the globe by a well-displayed dike of igneous rock that triumphantly cuts its way across the red Precambrian shale to form a gate to the white water below.

Immediately upon emerging from the wild Hance Rapid, it is a pleasure to stop and examine the two oldest formations of sedimentary rocks in the Grand Canyon Supergroup, and to dry off from the inevitable soaking received from the river. The gray, ledgy cliffs on either canyon wall belong to an ancient formation known as the Bass Limestone, which we already know contains traces of the most ancient life on earth in the form of algal stromatolites, or fossil algal mats. Although the age of these laminated rocks is not known for certain, it is probable that they are in the neighborhood of 1.5 billion years old; but still they show little metamorphic effects for their great antiquity.

It is inspiring to contemplate the significance of the past three days; journey. We have traversed geologic time from the age of dinosaurs in the rocks surrounding Lees Ferry backward into ages when shellfish and trilobites dominated the sea floors, as demonstrated by the Paleozoic beds. As we left the Tapeats Sandstone behind only yesterday, we entered the Late Precambrian ages, when the highest forms of life were soft-bodied wormlike organisms that defied preservation as fossils, and now we have come to the very earliest forms of known life ever to occur on our planet. The simple algae had few problems except to acquire sunlight with which to produce their own food supply and to remain where they would occasionally be bathed by seawater as the tides rose and fell across the primeval mud flats. Yet these simplest of plants were the ancestors to much of the proliferation of life we know today. We have seen the record of this explosion of the tree of life in reverse in only three days—from the largest forms of life ever to wander the landscape, to the simplest filamentous and unicellular green plants.

But we are not at our journey's end yet. Immediately beneath the Bass Limestone is a discontinuous layer of boulders known as the Hotauta Conglomerate, which was obviously eroded from very ancient and highly metamorphosed terrains. This "fossil" rubble marks the gates to the dark, foreboding inner gorges of the canyon, which contain some of the oldest rocks on earth: the Vishnu Schist. It is as if the depths of the earth were open to our gaze. It is like a journey to the earth's interior. It is as if one were regressing in time into the geologic dark ages, before life or oceans or atmospheres as we know them, as we enter the black depths of "Granite Gorge."

On August 14, 1869, Powell wrote: "We can see but a little way into the granite gorge, but it looks threatening.

"After breakfast we enter on the waves. At the very introduction, it inspires awe. The cañon is narrower than we have ever before seen it; the water is swifter; there are but few broken rocks in the channel; but the walls are set, on either side, with pinnacles and crags; and sharp, angular buttresses, bristling with wind and wave polished spires, extend far out into the river."

The sheer walls of metamorphic rock rise skyward until the inner gorge is a thousand feet deep, and the younger Precambrian and Paleozoic precipices and ledges have retreated and hidden from our view. The dark mood of the inner canyon is punctuated by the roar of the river as it forms several difficult rapids; Sockdolager, Grapevine, Clearwater, Zoroaster, and others. Our travels into the interior of the earth and the beginnings of geologic time seem at a climax as we near Bright Angel Creek and the suspension bridges that span the river, and find a myriad of granite dikes and sills that have almost completely assimilated the gneiss and schist. Surely the beginning of our very being is near at hand.

The climb up the canyon walls along the Kaibab Trail is anticlimactic, but it provides ample opportunity to recapitulate the geologic events we have witnessed and to reflect on the meanings of history and life and ancient environments now lost to the idiosyncrasies of time itself.

CHAPTER THREE

THE PLOT THICKENS

PALEOZOIC ROCKS OF THE CANYONLANDS

The Grand Canyon, although one of the best areas in which to see the Paleozoic rocks of the western United States, has no corner on the market. Most of the Paleozoic section can be traced upstream along the Colorado River throughout the length of Marble Canyon to Navajo Bridge. There, U.S. Highway 89 crosses the gorge near Lees Ferry, just before the strata dive beneath the surface and out of view at the Echo Cliffs monocline. This sharp fold in the crust separates Marble Canyon and its Paleozoic scenery from Glen Canyon and its Mesozoic strata a short distance downstream from Page, Arizona, and the Glen Canyon Dam. The Paleozoic rocks are buried beneath younger strata throughout the Glen Canyon region, only to reappear upstream where the Colorado River carves its course across another major area of arching called the Monument Upwarp. There another great episode of erosion by the Colorado River excavated the younger overlying beds to exhume the Paleozoic strata in a beautiful but rugged trench called Cataract Canyon. This upstream canyon was named by Powell because of fast, rough stretches of river lying within the chasm.

Cataract Canyon and its colorful tributary network form the heart of Canyonlands National Park. The depths of the canyons are virtually inaccessible by ordinary means and deter all modes of transportation except by air or rough-water river boats. Although the canyons do not reach the extreme depths of Grand Canyon, they are measured in thousands of feet, and the labyrinth of tributary canyons form a rugged and beautiful desert landscape second to none. The "canyonlands" are bounded in a general way by the eroded margins of the colorful sandstone and shale cliffs of the Mesozoic Era that once covered the region, only to be stripped away by the rampaging Colorado River system. This denudation exposed the Paleozoic layers throughout

THE GEOLOGIC TIME SCALE
FOR THE FOUR CORNERS REGION

ERA	PERIOD	MILLIONS OF YEARS AGO	FOUR CORNERS FORMATIONS
Cenozoic	Quaternary	0-1.6	soil, sand, gravel
	Tertiary	1.6-65	Terrace gravels, Diatremes Igneous intrusives
Mesozoic	Cretaceous	65-135	Mesaverde Group Mancos Shale Dakota Sandstone Cedar Mtn./Burro Canyon
	Jurassic	135-205	Morrison Formation Entrada Sandstone Carmel/Wanakah Formation Navajo Sandstone Kayenta Formation Moenave Formation Wingate Sandstone
	Triassic	205-250	Chinle Formation Moenkopi Formation
Paleozoic	Permian	250-290	Cutler Group: DeChelly Sandstone Organ Rock Shale Cedar Mesa Sandstone
	Pennsylvanian	290-325	Halgaito Shale Hermosa Group: Honaker Trail Fm. Paradox Formation
	(Rocks not exposed)		Pinkerton Trail Fm. Molas Formation
	Mississippian	325-355	Leadville/Redwall Fm.
	Devonian	355-410	Ouray Limestone Elbert Formation
	Silurian	410-438	Rocks missing
	Ordovician	438-510	Rocks missing
	Cambrian	510-570	Ignacio-Lynch Formations
Precambrian	Upper	570-2,500	Quartzite/granite
	Lower	2,500-4,500?	Metamorphic/granite

{*Dates in millions of years from Cowie and Bassett, 1989}

the inner canyon system, producing a geologic fantasia and scenic splendor that is truly unbelievable to the visitor.

Good roads lead to key overlook points along the periphery of the Mesozoic façade, where tantalizing previews of the magical inner canyons tempt the imagination. The true fantasy of Canyonlands is only realized by adventuring into the forbidding topographic maze by four-wheel-drive vehicles, horseback, river boat, or on foot. And then only under the direction of experienced local guides who know the ways and treacheries of the rugged desert.

Canyonlands doesn't exhibit the more ancient rocks, as does Grand Canyon, but displays magnificent exposures of sedimentary rocks deposited during the Pennsylvanian and Permian periods. Earlier strata are known to occur beneath these late Paleozoic beds because they have been penetrated by the drill at dozens of locations. Prior to the time the canyonlands became a national park, these deep holes were drilled in search of conditions within the sedimentary rocks that would be favorable for the formation and trapping of economical accumulations of oil and gas. Cuttings and cores were collected from the 4,000- to 10,000-foot-deep holes, and when examined by the geologist, revealed that the underlying Cambrian, Devonian, and Mississippian strata are similar in nature to those already described in the Grand Canyon and Marble Canyon exposures. Only a few wells have penetrated to the Precambrian basement, but these reveal a foundation of granite rather than metamorphic rocks beneath the region. Granite is a granular, crystalline rock formed by the cooling of molten rock materials, differing from lavas in that it cooled slowly and solidified deep within the earth rather than being rapidly cooled into fine-grained rocks upon extrusion at the earth's surface.

It is fascinating to attempt to reconstruct the history of the region from Precambrian times to the present from the drill holes and rock outcrops in the canyon walls. It is also important to reach back into these primeval times because the events of the dim, dark portions of history affect all subsequent developments in the evolution of the scenery, as we shall see. As explained in Chapter One, the interpretations are based on the doctrine of uniformitarianism, which states: "The present is the key to the past." The principle assumes that processes active at the earth's surface in modern environments can be expected to have been operative in the geologic past. In this chapter it will be demonstrated that the converse is also true—that the past is the key to the present. Events in the earliest geologic history controlled later sedimentation, which subsequently influenced the development of surface topography in and near Canyonlands.

Pre-Pennsylvanian Rocks

The early Paleozoic rocks of the Canyonlands are very similar indeed to the rocks previously described in Grand Canyon. This should come as no surprise, since they were deposited in similar environmental settings in the same seaways. Both regions lie along the shallow shelf that comprised the eastern margin of the Cordilleran seas of western Utah and Nevada during the early Paleozoic.

The great north-south Cordilleran seaway lay almost entirely west of the Colorado Plateau in Early Cambrian times, but as sea level began to rise relatively to the land level of that time, the shoreline with its beaches began its slow march eastward across Arizona and Utah toward the backbone of the continent. The wave action along the shoreline picked up sand formed by the weathering of Precambrian metamorphic and granitic terrains, and incorporated the loose sediment into its beaches. The beaches moved slowly eastward across what is now Grand Canyon, depositing beach and shallow marine sands that solidified into the Tapeats Sandstone. The shoreline passed the eastern limits of Grand Canyon country by the beginning of Late Cambrian time, and continued its march across Canyonlands in the closing millennia of the Late Cambrian. The beaches reached the mountain region of southwestern Colorado only at the close of the period. The Tapeats Sandstone, its offshore mud accumulations of the Bright Angel Shale, and the Muav Limestone, crossed Canyonlands country somewhat later than in Grand Canyon. The deposits, however, are almost indistinguishable in the two widely separated regions. The sea apparently withdrew westward at about the close of Cambrian time, for the dolomitized limestones at the top of the Cambrian interval suggest shallowing waters and development of tidal-flat environments across the entire region. As in Grand Canyon and the rest of the Colorado Plateau, no rocks of Ordovician, Silurian, or Early Devonian age are known to have been deposited in the Canyonlands region.

No geologic record exists in this region for some 140 million years, as strata of Late Devonian age are found to directly overlie the Cambrian beds. These Devonian rocks are largely composed of dolomite, a rock formed by an alteration in crystal structure of limestone by the addition of magnesium, probably occurring shortly after the deposition of the original lime sediments. The strata appear to have been deposited in shallow waters. Stromatolites and salt casts are occasionally found in them, and fossilized fragments of primitive fish have been collected from well cores. The Devonian section is called the Elbert Formation in this region. The formation includes some

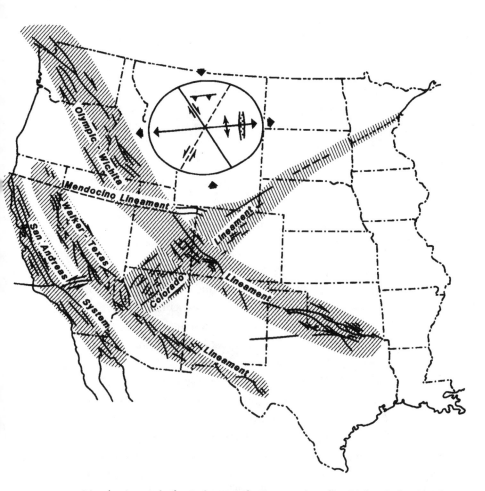

Map showing trends of major basement fracture zones that affected Paleozoic depositional patterns and later structures in the western United States (Baars and Stevenson, 1981).

thin beds of a bright apple-green shale and local sand-bar-like sandstone lenses that form a reservoir rock for oil at Lisbon Valley, near Moab, Utah.

The shallow-water dolomite and sandstone of the Elbert Formation give way upward to an extensive bed of dark brown limestone known as the Ouray Limestone. The Ouray was deposited in open marine conditions at a time when the Devonian sea was at its maximum invasion onto the continent. It contains numerous brachiopods, crinoids, and a few "forams," which attest to moderate ocean depths and conditions of normal, unrestricted marine circulation. The fossils dictate that sedimentation of the Ouray Limestone

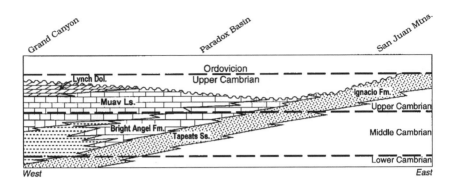

Generalized cross section showing how Cambrian formations become younger moving from the Grand Canyon eastward to the San Juan Mountains.

occurred at about the close of Devonian time or the beginning of Mississippian time, and may have been deposited across this important temporal boundary. The rocks of the Elbert Formation resemble the Temple Butte section in Grand Canyon to some extent, but are much more uniform and extensive in occurrence. The Ouray has no known counterpart in Grand Canyon, and may have been removed by pre-Mississippian erosion in that region.

A relatively brief episode of erosion occurred following the deposition of the Ouray Limestone—at least in some areas, such as central Colorado and the Grand Canyon. In canyonlands there is really no convincing evidence that sedimentation was interrupted at all between Ouray and Redwall times, for the limestone and dolomite of the Redwall appear to grade downward into the Devonian strata. The Lower and Middle Mississippian beds are almost identical in rock type and fossils to the Redwall of the Grand Canyon, so that it has recently been suggested that the name "Redwall" be used in the Canyonlands and Four Corners region. Indeed, even the prominent chert member in the Grand Canyon can be traced through outcrops and deep wells into this region. The erosional break in the middle of the formation can be found throughout the Colorado Plateau. Thus the conditions of deposition must have remained rather constant across the entire Colorado Plateau, and even beyond, during Mississippian time. The Redwall Formation in Canyonlands consists of a lower part made up of sugary-textured dolomite and chert, and an upper portion of dolomite and limestone that varies a great deal in rock type and fossil content depending on local conditions at the time of deposition. The two members are separated by a thin layer of limestone pebbles that formed on a widespread surface of erosion within the formation.

The formation is tantalizing to petroleum geologist because the dolomite layers are almost always quite porous, but most of the time the porosity contains salt water rather than oil or gas. The valuable petroleum products do occur locally in the Redwall Formation in and near Canyonlands, at Lisbon Valley, Big Flat, and Salt Wash fields. Mounds of crinoid debris and lime mud accumulated in the shallow Mississippian sea at those localities. The sediment build-ups were later recrystallized to form the necessary porosity to produce sufficient petroleum reserves of economic significant.

Petroleum does not occur in caverns or pools within the rocks, as many people believe, but instead inhabits the very small pores between sand grains in sandstone and between crystals or fossils in limestone and dolomite. The oil forms when changes are imposed upon organic materials deposited within

Close-up photograph of a bedding-plane surface of a dolomite layer from the upper Elbert Formation in the San Juan Mountains of southwest Colorado. The raised cube-shaped structures are thought to have been formed by salt cubes growing beneath the surface of tidal-flat muds as the waters near the surface of the sediments evaporated. They have now been replaced by dolomite, so they are called "salt casts." Background grid is 1/10th inch squares.

Westwater Canyon where the Colorado River crosses the crest of the Uncompahgre Uplift of the Colorado-Utah border region. The lower cliffs are metamorphic rocks of Precambrian age, overlain by the Triassic Chinle and Jurassic Wingate formations.

the original sediment. The process requires extensive intervals of geologic time under rather high temperature and pressure occurring within the crust of the earth. The oil must then be concentrated into porous reservoir rocks from which it cannot escape before it can be unleashed and utilized by man. In the case of the Redwall Formation in Canyonlands, this is accomplished by the migration of petroleum into the fossil crinoid-mud banks. Because these limestone build-ups are fairly numerous in this region, the Redwall is a major potential source of natural hydrocarbons and constitutes a major objective in the search for new reserves.

Meanwhile, during the deposition of these sediments in the shallow marine waters of the Colorado Plateau country, structural forces were at work molding fracture patterns that were to dominate the region for eons to come. It has been shown that several fault blocks were produced in late Precambrian time in the vicinity of Grand Canyon. The elongate fault-bounded ridges were buried by Cambrian sedimentation, and their subsequent effect on later deposition was thus minimized. Similar northwest-trending faults originated in the eastern reaches of the Canyonlands province. Prior to the deposition of the Cambrian Tapeats Sandstone, elongate fault-bounded ridges formed that acted as low islands or submarine rises in the Cambrian sea. Deep wells drilled along these ancient features have shown that the individual formations of the Cambrian section thin markedly over the fault blocks, confirming the interpretation. Intermittent periods of exposure to the erosional elements produced local unconformities along the crests of the fault blocks within the Cambrian interval, as revealed in cores from the deep probes.

Unlike the situation in Grand Canyon, the faults were intermittently rejuvenated during the remainder of the early Paleozoic, influencing sedimentation of the Devonian and Mississippian sequences. Sand bars, which develop only in relatively shallow waters due to the effects of breaking waves or tidal current movements, developed along the high flanks of the ancient fault blocks in Devonian time; the related dolomite beds thin over the structures. Similarly, the crinoid-mud banks of the Redwall Formation, which contain the lion's share of Canyonlands oil and gas, were situated on the shoaling submarine topography along the fault blocks. Similar sediment banks in modern lime-depositing environments in south Florida develop only in very shallow waters.

Thus, it is probable that the long-continuing submarine topography formed by the recurring fault movements was responsible for the favorable conditions that led to the construction in and near Canyonlands of reservoir-quality rocks. Today, these are the rocks that contain commercial quantities of oil and gas. Structure has been influential on subsequent sedimentation in this case, thereby indirectly controlling the position of potential reservoir rocks. Keep these fault blocks in mind, for we shall see later that they continued to be important in the geologic development of the region and the closely related scenery that resulted.

Life and Death of a Salt Basin—the Paradox Basin Story

The legacy left to the Pennsylvanian Period is a series of comparatively uniform layers of marine rocks, broken locally but systematically by a swarm of northwesterly-trending fractures in the eastern portions of Canyonlands. The general pattern of the early Paleozoic seaways and basins was to change radically in Pennsylvanian time. The major seaways and shelves gave way to the development of local but deeper basins with sharply contrasting intervening uplifts in late Paleozoic times. The cause of this major revision of pattern remains something of a mystery, as advocates of plate tectonics theory speculate that a major collision of South America with North America was the culprit. The significant fact remains that the basins and mountain ranges came abruptly into existence in mid-Pennsylvanian time throughout the western United States. In the vicinity of Canyonlands, an elongate northwesterly-trending basin formed along the earlier fracture system, producing a depression in the crust of the earth that measured some two hundred miles in length. In all, the pre-Pennsylvanian strata were depressed some twenty thousand feet into the crust of the earth. This major invagination in the earth's crust is known as the Paradox basin.

The basin is bounded along the northeast margin by a compensatory uplift of a major mountain range. The upland was to dominate the landscape of the region for more than two geologic periods of time. The highland supplied untold billions of cubic yards of sediment to the basin, as erosion attacked the newly formed uplift with a vengeance. The major ancient mountain range uplifted along the earlier fracture system, becoming in Pennsylvanian and Permian times a fault-block range with perhaps several thousand feet of relief. It is known today as the Uncompahgre uplift segment of the Ancestral Rocky Mountains, for its position approximates that of the present-day Uncompahgre Plateau in eastern Utah and western Colorado. Let us see what effects these seemingly catastrophic events had on the depositional history of the region.

The oldest exposed strata in Canyonlands National Park are rock layers of Middle Pennsylvanian age in the depths of Cataract Canyon. They were deposited in the rapidly sagging Paradox basin. The deepest excavations by the Colorado River occur at the mouth of Gypsum Canyon, named because of a thick sequence of bedded gypsum exposed by erosion in the lower three hundred feet of the chasm.

Gypsum is a soft, white rock composed of hydrous calcium sulphate. It forms naturally when seawater is highly concentrated by rapid evaporation,

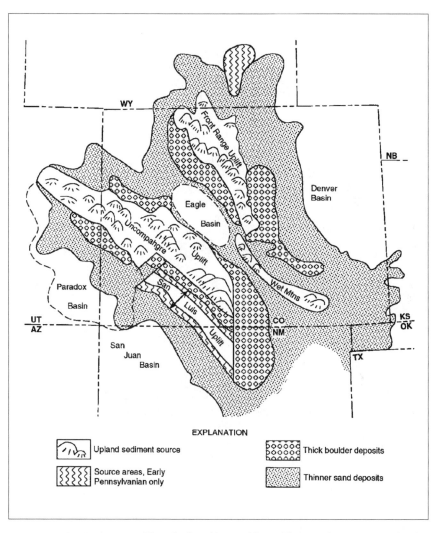

Map showing locations of the uplands and basins of the Middle Pennsylvanian Ancestral Rocky Mountains. Both the Paradox and Eagle basins received thick evaporite deposits at this time.

usually in a highly arid climate, until crystals begin to precipitate from the supersaturated solution. The highly saline waters need only be concentrated a little more until halite (salt) begins to crystallize, and consequently halite and gypsum commonly occur together in alternating layers. The kind of assemblage of rocks formed by the rapid evaporation of seawater is called "evaporite" by geologists. Of course, halite is exposed at the surface only in extremely arid regions, for the highly soluble mineral is readily dissolved by

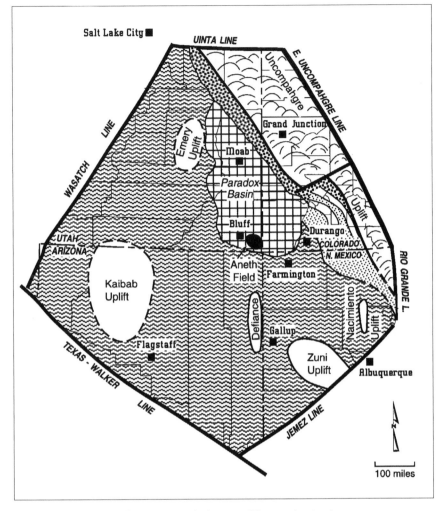

Map showing distribution of sedimentary rocks during Middle Pennsylvanian time.
Wavy pattern indicates shallow marine deposits; cross pattern shows area of evaporite beds;
dot pattern indicates area of sandstone deposition; coarse dots show area
of conglomeratic deposits; and blank areas are locations of non-deposition.

surface and near-surface fresh waters. This explains why there are no salt
outcrops in Canyonlands.

Other occurrences of gypsum are known in Cataract Canyon, specifically at Spanish Bottom, near the confluence of the Green and Colorado
rivers and at two other locations a short distance downstream. These bodies
of evaporite are not in their original bedded configuration, however, but
have been squeezed upward through the overlying layers in plastic intrusives

THE COLORADO PLATEAU

much like toothpaste. Much-larger evaporite intrusions underlie several of the elongate valleys east of Canyonlands in spectacular geological displays. We will return shortly to discuss how these interesting structures formed.

Let us digress for a moment to set the stage by looking into the framework of the Paradox basin through our geologic crystal ball. Wells drilled throughout the region reveal that several major geologic events preceded the deposition of the evaporites in the Paradox basin. Following the deposition of the Redwall Limestone in Middle Mississippian time, the seas withdrew from the Rocky Mountain and Colorado Plateau regions, exposing the newly formed limestone to weathering. It is not known exactly when this interruption occurred, but is was during all or part of the Late Mississippian to Early Pennsylvanian time interval, for there are no rocks of that age in the region. Weathering of the limestone terrain caused a layer of brick-red soil to form over most of the western interior of the continent. Cavernous underground passageways for the movement of rain and ground waters formed, and sinkholes pockmarked the terrain in some areas. The results of this period of extended weathering are summarized in the red mudstone layers of fossil soil and the red soil-filled fractures and caverns in the Redwall Limestone.

The red, purple, and maroon shale or mudstone developed on the weathered surface separate the Mississippian limestone from the overlying strata throughout most of the intermountain west. This shale unit is especially well developed in the Paradox basin country. Although the reddish interval is usually only a few feet thick, it is significant because it tells the story of extensive soil development. The beds are called the Molas Formation, for exposures near Molas Lake and Molas Pass in the San Juan Mountains of southwestern Colorado. The formation name has been used throughout the western states wherever the soil mantle has been recognized. The undisturbed ancient soil forms only a portion of the Molas Formation, however, as the soil was locally reworked and redistributed by streams and the earliest invasion of the Pennsylvanian seas. Ancient stream channels filled with sand and gravel are often found within the mudstone. The pebbles usually are composed of limestone containing Redwall fossils, indicating that they were derived from the extensive underlying formation.

This period of soil formation and reworking by streams was interrupted in Early Pennsylvanian time when the sea again covered the region now occupied by the Paradox basin and Canyonlands. At first the incursion of the sea redistributed the upper parts of the red oil debris into uniform layers that today contain fossils of the marine animals that lived in the shallow sea. In

Gypsum intrusive plug at the mouth of Lower Red Lake Canyon near head of Cataract Canyon, Canyonlands National Park, Utah. Gypsum (in foreground) is from the Middle Pennsylvanian Paradox Formation. The cliffs are in the Pennsylvanian Honaker Trail Formation and Elephant Canyon, the latest Pennsylvanian; and Cedar Mesa formation (Permian) in ascending order.

places the soil material was completely removed, only to be redeposited in deeper, quieter areas of the sea floor nearby. The result was an erratic distribution of the red mudstone that symbolizes the irregularities of the sea-floor topography. The Molas Formation also acts as a punctuation mark in the record, because of its distinctive appearance.

As the reworking of the red soils was completed by the advancing sea in Early Pennsylvanian time, lime sediments began to dominate the underwater scene, forming relatively thin layers of marine limestone that extend throughout large areas of the Paradox basin. The fossiliferous strata have come to be known as the Pinkerton Trail Formation, marking the complete inundation of the Paradox basin region by Middle Pennsylvanian time.

As deposition of limestone of the Pinkerton Trail progressed, the central portion of the Colorado Plateau region began to sag along the lines of ancient fracture systems in the vicinity of Moab, Utah. The gentle initial depression of the seafloor marked the beginning stages of the great Paradox basin and the initiation of salt and gypsum sedimentation in the deeper, more stagnant structural troughs. Salt deposition rapidly filled the elongate fault-formed submarine troughs and spread outward as the region continued to sag, until at its maximum development the oval outline of the salt-depositing basin extended as far south as northern Arizona and New Mexico and as far north as Price, Utah. Canyonlands and the surrounding areas for many tens of miles were completely covered with hundreds to thousands of feet of salt and gypsum.

It is believed that the very thick evaporite deposits resulted from the stagnation of the sea when the region became structurally depressed into a deep basin. The climate must have been extremely hot and dry to permit the evaporation of vast amounts of seawater. Even then, large amounts of fresh seawater must have been periodically or continuously added to the basin, for no single basinful of seawater could contain enough salt to form the 6000 or 7000 feet of evaporites that eventually came to rest in the Paradox basin. It is probable that seawater continued to enter the depression across the shallow margins of the basin, only to be highly concentrated to a brine by the intense evaporation. As the brine became sufficiently dense, it would sink to the bottom of the rapidly subsiding basin and stagnate rather than be flushed across its shallow periphery. Thus the fresh seawater entering the basin across the shallow threshold acted as a dynamic barrier that restricted the escape of the dense bottom waters. As the trapped waters became sufficiently saline to be supersaturated with salts, precipitation of halite and/or gypsum commenced and bedded evaporites resulted. By this process, fresh supplies of salt were being added to the basin daily, permitting the great thicknesses of evaporite to accumulate. Thin layers of black shale that are high in organic debris alternate with the evaporite beds, forming a cyclic alternation of rock types throughout the section. The intermittent supply of mud may have been regulated by fluctuations in sea level that occurred throughout the world during Pennsylvanian time. Thin black shale layers occur at vertical intervals of about 100 to 300 feet and appear to be laterally extensive throughout most of the basin. The dying phases of this evaporite episode are represented by the gypsum layers exposed in the depths of Cataract Canyon, at the mouth of Gypsum Canyon.

While the interior of the Paradox basin was receiving evaporite sediments, its shallow margins were being bathed with relatively fresh seawater as it entered the basin. These shelf areas on the south and west sides were the favored habitats for abundant marine life, and consequently limestone was deposited more or less simultaneously with the salt beds in the basin interior. The limestone sections are much thinner than their contemporary salt deposits within the basin, being only a few hundred feet in thickness compared to a few thousand feet of the salt.

The shallower waters hosted an unusually large number of algae, along with the normal complement of marine animals. The green plants were of a type that secreted calcium carbonate ("lime") within their tissues. Upon the death of the plants, calcareous algal particles were formed that contributed vast quantities of lime sediment to the sea floor. The calcareous algae, like their modern counterparts in tropical seas of south Florida and the Bahamas, liked conditions best in very shallow, warm water that is very quiet. Because of this preference for shallow water, they grew in abundance on local shallow shoals. The concentration of algal growth produced accumulations of calcareous particles that locally thickened the deposits and intensified the extreme shallowness of the bank. Thus, the preference for a particular environment by the algae perpetuated the local condition and formed mounds of algal debris that attained considerable thickness over periods of a few million years.

The details of the texture of the sediments was rather unusual because of the nature of the algal material. The algae were apparently about the size and shape of a small head of lettuce, except that the individual "leaves" were not so tightly infolded. When the plant died, calcareous plates much like rock-hard potato chips fell from the algal thallus to become fossil fragments. Because the leaflike debris was thin and irregular in shape, the deposits were composed of irregular platy chips and large intervening pores, forming a highly porous and permeable sediment. The mounds of porous fossil algal material formed locally thickened pods of limestone that are now called "*bioherms.*" Sometime after the burial of the algal bioherms, petroleum, generated from the highly organic black shale beds of the Paradox basin, migrated through the rocks with the ground waters until it found its way into the porous limestone traps. Exploration drilling in the past few years has discovered the bioherms, and millions of barrels of oil have been produced from them in the Four Corners region of the southern margin of the Paradox basin.

Meanwhile, on the northeastern flank of the rapidly subsiding Paradox basin, the large mountain range known as the Uncompahgre uplift was rising rapidly alongside the trough. The side of the range adjacent to the basin was bounded by enormous fractures along which the uplift occurred, forming a series of northwest-trending faults that bounded the uplifted block. The fault-block mountains developed along the same swarm of fractures that are now seen to underlie the Paradox basin. The overall structural grain and

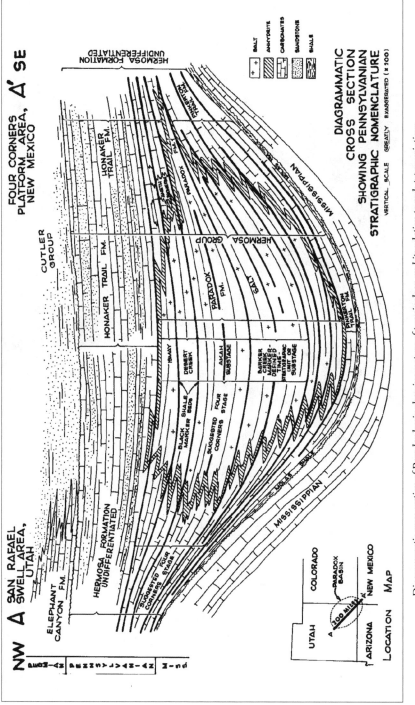

Diagrammatic cross section of Paradox basin showing nature of terminology and its relation to rock-type variations.

The Colorado River in Cataract Canyon in the heart of Canyonlands National Park. The lower ledgy cliffs are in the Pennsylvanian Honaker Trail and Elephant Canyon formations; the upper massive cliffs are the Permian Cedar Mesa Sandstone.

geologic history of the uplift and the basin are closely related.

That the major mountain range was actually present and began its rise in Middle Pennsylvanian time is known from several lines of evidence. In the first place, the mountain chain lies along the Uncompahgre Plateau, that consists of metamorphic rocks of Precambrian age. The basement rocks are covered by sedimentary rocks of Triassic and Jurassic age, as seen in Colorado National Monument and Black Canyon National Park. No strata of Paleozoic age are present on the plateau, although it lies immediately adjacent to the very thick, but topographically lower, Paleozoic deposits of the Paradox basin.

Drilling in the basin alongside the Uncompahgre Plateau has revealed accumulations of 9,000 to 12,000 foot thicknesses of course sand, pebbles, and boulders of quartz and feldspar. The sediments were derived from the Precambrian granitic and metamorphic rocks of which the plateau is composed. This indicates that great quantities of the ancient rock were removed from the region of the plateau sometime in the geologic past. Sandstone composed mainly of quartz and feldspar grains are called "arkose" and comprise the deposits of the eastern Paradox basin. The thick arkose becomes finer in grain size southwesterly into the basin and away from the uplift, strongly suggesting that the source area for the sand and gravel was the Uncompahgre region. They interfinger and otherwise grade into salt, gypsum, shale, and limestone of Middle Pennsylvanian age, thereby dating not only the arkose beds but also the time of uplift of the source area—the Uncompahgre uplift. The fault-block mountains began their rise in Early to Middle Pennsylvanian time, as the Paradox salt basin was subsiding, and, from the amount of arkosic debris present, they must have risen to considerable heights to have supplied such unbelievable quantities of sand and gravel.

Salt-Intruded Anticlines

As already mentioned, a series of smaller faults parallel the Uncompahgre uplift in the depths of the Paradox basin. They divide the deepest part of the basin into a series of elongate ridges and troughs that were deeply buried under several thousand feet of evaporites. The oldest beds of salt in the basin are found in these down-faulted trenches which underlie the region between Canyonlands National Park and the Uncompahgre uplift. When these structural features had been inundated with salt in Middle Pennsylvanian time, the uplift of the Uncompahgre mountains had begun, and clastic sediments from the foreboding range were being deposited along the margin of the salt basin.

The weight of overburden of these arkosic sediments was great and becoming more unbearable as time went on and sediment thickness increased. The salt began to flow as a plastic mass, caused by the increasing weight of the arkose, and it flowed toward the southwest—away from the region of greatest overburden. When the great masses of flowing salt encountered the

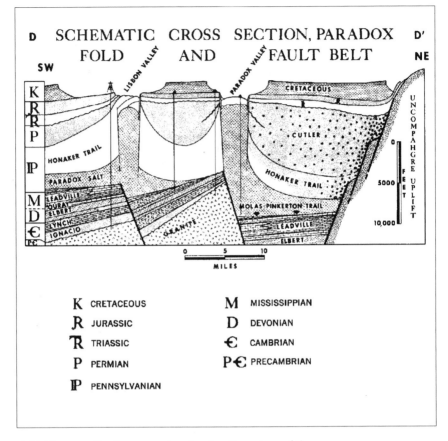

SCHEMATIC CROSS SECTION, PARADOX FOLD AND FAULT BELT

K	CRETACEOUS	M	MISSISSIPPIAN
R	JURASSIC	D	DEVONIAN
ᚱ	TRIASSIC	Ꞓ	CAMBRIAN
P	PERMIAN	PꞒ	PRECAMBRIAN
ℙ	PENNSYLVANIAN		

A highly generalized cross section showing the nature of the salt-intruded anticlines at depth. Each salt wall is underlain at depths exceeding 10,000 feet by the ancient faults that trend northwesterly, parallel to the Uncompahgre uplift. The fault blocks created buttresses that deflected the westward-flowing salt into upward-flowing salt wedges that penetrated the overlying strata. The thick Cutler arkoses lying against the front of the uplift caused the salt to begin flowing by creating a localized overburden that loaded the salt with unequal stresses. The salt-intruded anticlines become smaller toward the southwest as the salt source beds thin and the Cutler overburden decreases abruptly.

The town of Moab, Utah, nestled into Moab Valley, a breached salt-pierced anticline.
The La Sal Mountains form a striking background.

fault-block ridges in the bottom of the Paradox basin, the viscous mass of salt was deflected upward along the faults to form vertical walls of salt which penetrated the overlying strata. As the weight of the arkosic debris increased, the salt continued to flow until the bulk of it was contained in the intrusive salt walls. Salt flowage pierced all overlying strata, following the paths of least resistance to areas of lower pressure of overburden.

This sequence of events was not as catastrophic as it may sound, for the process apparently took the better part of 150 million years. Salt flowage began in about Middle Pennsylvanian time and is known to have progressed through all of Permian, Triassic, and Jurassic times. Sediments deposited during those intervals either thin over the salt bulges, or were penetrated and upturned along the salt-piercement structures. Movement of the salt ceased in Late Cretaceous time, when the first sedimentary strata cross the structures without serious consequences. The contorted beds of salt and gypsum extend downward to a depth of about 15,000 feet, as revealed by deep wells drilled into the salt structures.

The salt may have actually reached the surface at times during the process of growth by salt intrusion, but there is little evidence to support the idea. However, the intruded evaporites presently lie at or near the surface along the crests of the structures. The climate is not dry enough for the salt itself to actually crop out, but beds of gypsum and black shale that are highly contorted occur along many of the surface features. Bedded salt has been penetrated by the drill at depths of only about five hundred feet in many cases, beneath the concentrated gypsum caprock.

Ground waters moving through the near-surface rocks have encountered the salt masses in fairly recent times, and have dissolved the salt from the upper parts of the structures, leaving the less-soluble gypsum behind. The result has been a reduction of volume of salt near the surface, and overlying strata have collapsed into the elongate crests of the salt features. The present surface expression of these collapsed salt structures is extensive elongate valleys. These "salt valleys," as they are called, are not little features by any means. Moab Valley, in which the town of Moab, Utah, is located, is about 17 miles long; Salt Valley, in and near Arches National Park, is about 23 miles long; and Paradox Valley, the largest of the lot, is about 35 miles in length. These are only three of nine major salt valleys in the region.

The genesis of the term "Paradox" can now be considered, with the nature of the Paradox basin and the salt-intruded structures in mind. Early pioneers in the region noticed that the northwesterly-trending valleys were very prominent topographic features. For some unknown reason the major rivers did not flow along the valleys, as is customary. Instead, such large rivers as the Colorado and the Dolores cut sharply across the valleys in strange fashion. The Colorado River crosses Moab Valley near the town of Moab at nearly right angles, and the Dolores River was found to do the same unusual trick in another valley southeast of Moab. This incongruity caused the pioneers to name the latter valley "Paradox Valley" for the paradoxical geomorphic phenomenon. Early geologists in the region saw that Paradox Valley contained rather extensive exposures of contorted gypsum, which they recognized as being of salt-flowage origin, and that the valley resulted from the denudation of a salt-piercement structure. At that time they didn't know exactly where the salt had come up from, but believed correctly that it must be of Pennsylvanian age. They proceeded to give the evaporites the name "Paradox Formation" for these occurrences in Paradox Valley. It follows naturally that when the nature of the salt basin became known through deep drilling it was called the Paradox basin.

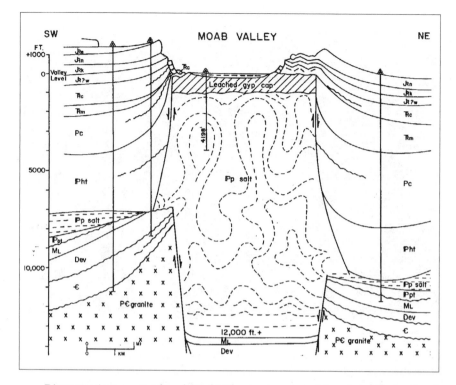

SW   MOAB VALLEY   NE

Diagrammatic cross section through Moab Valley, showing nature of the rocks in the subsurface of the salt-penetrated structure. Vertical lines indicate the locations and penetrations of key deep wells. Other salt-intruded structures in the area, such as Salt Valley, Castle Valley and Paradox Valley, have similar characteristics. Note vertical scale at left.

Upheaval Dome

Several small, circular salt-piercement structures occur in the region, but they are relatively insignificant when compared to the spectacular salt valleys. The gypsum exposures in Cataract Canyon at both Spanish Bottom and Cross Canyon are of this variety.

One large circular feature in Canyonlands National Park known as Upheaval Dome may also be of this nature. A salt core is not exposed there, but the overall shape makes Upheaval Dome a good candidate for a salt-intruded origin. Several mysterious features cloud the true nature of the spectacular dome, making it difficult to evaluate without drilling the structure. Although the surface feature has all the characteristics of a good salt dome, i.e., sharply upturned strata, a surrounding depression or "rim syncline," etc., geophysical

Upheaval Dome in northern Canyonlands National Park, Utah. The circular pattern of sharply upturned strata are the Navajo-Kayenta-Wingate formations of Jurassic age. The centrally located small pinnacles are splinters and sandstone dikes that have penetrated the Moenkopi Shale from the White Rim Sandstone below. Many geologists believe that the sharp structure is a classic example of a salt dome. The late astrogeologist, Eugene Shoemaker, argued that the dome was created by a meteoric impact that occurred some 65 million years ago.

studies have shown that the rocks underlying the dome are not deficient in mass, as salt should be. This has led some geologists to believe that an intrusive igneous mass domed up the structure instead of salt. The same geophysical evidence, however, does not indicate the presence of such a dense rock at depth. Some geologists have postulated that the dome was originally formed by salt intrusion in Pennsylvanian through Jurassic times and, much later, perhaps in Tertiary time, partially displaced by igneous intrusions at depth. Surprisingly, another sister plug was noted by the same geophysical surveys to lie immediately southeast of Upheaval Dome, but there is no surface expression of the intrusive body and it had previously gone undetected.

In recent years, specialists in lunar geology, looking in vain for meteorite impact features everywhere on earth, have concluded that Upheaval Dome is indeed an impact feature. The late Eugene Shoemaker, the "Father of Lunar Geology," postulated that a meteorite (bolide) struck the earth here some 65 million years ago. The shock of the impact formed the deep fracturing and rebound dome we see today, and the resulting meteoritic debris and upper disturbed sedimentary rocks having been removed by recent erosion. He restudied the structure in the field and found evidence to support the theory. A proper scientific conclusion is that "we find what we look for."

More recently, a team of geologists from Exxon Production Research Company and the Bureau of Economic Geology, University of Texas, again mapped Upheaval Dome in great detail. Published in late 1998 in the Bulletin of the Geological Society of America, the study concluded that the feature was indeed formed by salt penetration from depth. They presented convincing evidence of salt movement having occurred throughout Pennsylvanian through Jurassic time, rather than having been formed instantaneously by impact. They further concluded that as the salt intruded all overlying rocks, and perhaps flowed onto the surface, the salt blob was pinched off from the source beds at depth as the supply of salt was exhausted. The pinched-off salt body, and all overlying deformed strata, was later eroded from the site, as well as the surrounding plateau country. While this theory may sound far-fetched, there are numerous examples of pinched-off salt domes in the Gulf Coast of the United States and elsewhere throughout the world, verified by geophysical studies and deep drilling.

Both explanations for the origin of Upheaval Dome, meteorite impact or salt piercement, rely on evidence now missing due to erosion. Because of this, both conclusions may be suspect to some geologists. The evidence amassed to reveal continued growth over several geologic periods, however, turns the tide in favor of salt penetration from the bottom up. Afterall, we know there is sufficient salt present in the area to form salt-intruded structures, and there are several undisputed examples just a short distance to the east. That the salt blob was pinched off at depth may be doubtful to some, this explanation is by far the most plausible explanation presented to date. It certainly fits the details of the surface geology. For now, at least, it is a pinched-off salt dome.

Salt-Related Collapse Features

Other spectacular features in Canyonlands owe their existence to the presence of salt in the Paradox Formation. Of major significance is the jumbled rock "fantasyland" in southern Canyonlands National Park called the "Needles District." There, the maze of rock pinnacles and spires were carved from a sandstone of Permian age, the Cedar Mesa Sandstone, by erosion along narrow, linear up-thrown fault blocks, separated by troughs or elongate valleys of the intervening down-thrown fault blocks. The resulting topography is that of elongate "coxcombs" of intricate rock needles and intervening "racetrack" valleys that permit north-south travel by wheeled vehicles. Travel is relatively easy in this region provided one does not wish to cross the fault blocks.

The origin of the so-called "Needles fault zone" and "Grabens" is closely related to the underlying Paradox salt and its subsequent solution by ground waters. The intricately faulted region lies across the structural nose of the large up-arched feature known as the Monument Upwarp, where it plunges northward toward the confluence of the Green and Colorado rivers. The bedded Paradox salt underlies the region at a depth of only about two thousand feet, where actively moving ground waters are still effective. It is probable that the underground waters that drain the Monument Upwarp and its relatively high country, flow down the dipping strata toward Cataract Canyon to the north and west, and remove some of the upper salt beds by dissolution. As in the formation of the collapsed crests of the salt-flowage structures, the solution of salt beneath the Needles country reduced the volume of rock beneath the surface, creating rather extensive voids within the strata. The collapse of the overlying beds would be imminent in this event, and would surely form large scale fracturing and complicated faulting, as we see at the surface of the Needles and Grabens area today. The large fractures and faults can be seen to extend in depth to the top of the Paradox evaporite section along the walls of Cataract Canyon, but they do not cross the Colorado River to the western canyon walls. This appears to be evidence that the fracturing terminates at the river, where the groundwater flow is diverted along the drainage paths established by the Colorado. The linear down-dropped valleys are today known as "The Grabens." In the Grabens area, the collapsed fault blocks are gliding down-dip toward the canyon. The Needles District, then, is comprised of spectacular scenic wonders that were formed by unusual interrelated circumstances of both physical and historical geology.

The Honaker Trail Formation

The upper third of Pennsylvanian time witnessed the return to normal, open marine conditions in the Canyonlands region. Fossiliferous marine limestone beds, with interbedded marine sandstone and shale, blanketed the earlier salt basin with a cover of at least fifteen hundred feet of strata now known as the "Honaker Trail Formation." These beds of tropical marine sedimentary deposits form the gray inner gorge of Cataract Canyon with its prominent multilayered topographic expression. They comprise the characteristic vertical, impenetrable walls of the inner canyon, which may be punctuated by steep slopes of eroded shale beds where the canyon widens. It is the Honaker Trail Formation that culminates the ruggedness and scenic grandeur of Canyonlands by hiding the canyon-carving Colorado River in impassable and mysterious gray depths through the heart of the park.

Honaker Trail time closed in Canyonlands country with a shallowing and subsequent withdrawal of the sea. The actual break is not particularly obvious in most places. The time boundary is usually represented by an abrupt change from marine limestone of the Honaker Trail Formation to red beds of the overlying Halgaito Shale, formerly believed to be of Permian age. When the unconformity is studied on a regional basis, however, it becomes an obvious break in the geologic record and reveals important relationships. The most drastic change is that the seaway shifted significantly toward the northwest, toward the rapidly subsiding Oquirrh seaway in the present vicinity of Salt Lake City and Provo, Utah.

If we study the age of the latest Honaker Trail strata that directly underlie the younger rocks, we find a definite pattern of regional changes. At Moab the coarse-grained Cutler Formation directly overlies latest Pennsylvanian marine beds, indicating that little or no strata have been removed by pre-Permian erosion. In the vicinity of the confluence of the Green and the Colorado rivers, in the heart of Canyonlands National Park, the younger rocks directly overlie beds in the Honaker Trail that are Late Pennsylvanian in age but are far older than latest Pennsylvanian, as at Moab. Also, local angular relationships are present at some localities, such as near the confluence of the Green and Colorado Rivers, suggesting that a slight tilting of the Honaker Trail strata occurred prior to the erosion and subsequent deposition of the younger sediments of latest Pennsylvanian (Virgilian and Bursumian) age. The tilting was no doubt a function of continued early growth on the Meander anticline, a northeast-trending salt-intruded structure along which the Colorado River flows.

West of the Green River, deep wells have found the latest Pennsylvanian rocks resting directly on Middle Pennsylvanian strata, and farther northwest, on what is called the San Rafael Swell, Permian strata overlie rocks of Mississippian age. In that region the entire Pennsylvanian section was removed by erosion prior to Permian time. This was the culmination of a broad Late Pennsylvanian uplift now know as the Emery uplift, which lay at the present-day site of the San Rafael Swell. So the mid-Pennsylvanian was brought to a close in the Canyonlands country by gentle uplift, withdrawal of the sea, and erosion of the lowland areas.

Near the confluence of the Green and Colorado Rivers, rocks that overlie the Honaker Trail Formation are quite similar to the Honaker Trail Formation. However, they were deposited in a different seaway. The region in which the Paradox and Honaker Trail Formations were deposited, was now a coastal lowland, as the region of marine deposition had moved to the northwest in an arm of the Oquirrh sea. The largely marine sequence, named the Elephant Canyon Formation for exposures near the Confluence, was originally believed to be of Early Permian (Wolfcampian) age. More recent studies of the fusulinids found in the rocks reveal that the age is not Permian, but latest Pennsylvanian age.

Fusulinids are small, but highly complex fossil Foraminifera that evolved very rapidly during Pennsylvanian and Permian time. For that reason, they are used universally to date and correlate rocks of late Paleozoic age. As studies progressed over the years, the fusulinid species found in the Elephant Canyon Formation were found to be somewhat older than first suspected. This made little difference in geologic interpretations in the region, except that the Elephant Canyon Formation is seen to be latest Pennsylvanian, rather than earliest Permian in age.

And then there is the problem of establishing a definite Pennsylvanian-Permian boundary on a global scale. With the final establishment of a definite lower Permian boundary by Russian paleontologists in the southern Ural Mountains of Russia and northern Kazakhstan, the lower Permian boundary as was customarily used in North America was necessarily moved up in the section. Thus, rocks we previously called Permian are now latest Pennsylvanian in age. The top of the previously recognized Pennsylvanian (the Virgilian Stage) is no longer the latest stage in the time period. Charles Ross and his wife June, have proposed that rocks of this nameless interval of time be called the Bursumian Stage, named for the Bursum Formation of southern New Mexico and West Texas that contains the fusulinid species in question.

Now, the lower part of the Elephant Canyon Formation is known to be Virgilian and Bursumian in age, and the upper part of the formation that occurs in the vicinity of the San Rafael Swell is of true Early Permian age (Wolfcampian). Because the Elephant Canyon Formation in Cataract Canyon interfingers southward with red beds, the equivalent Halgaito Shale is now seen to be latest Pennsylvanian in age. The upper Permian part of the Elephant Canyon interfingers with the Cedar Mesa Sandstone in the subsurface west of Cataract Canyon, so the Cedar Mesa Sandstone of eastern Utah is now considered to be basal Permian (Wolfcampian) in age. Whew!

Pedestal spires formed by erosion along closely spaced fractures near the White Rim Trail in Canyonlands National Park, Utah. The caprock is the Permian White Rim Sandstone overlying the softer upper Cutler (Organ Rock) siltstone beds.

Obituary of the Ancestral Rockies—the Permian System

Without a doubt the most spectacular and mystical scenery on the Colorado Plateau, or for that matter the world, resulted from the widespread exposure and sculpturing of rocks of Permian age by the relentless processes of desert erosion. The sedimentary rocks of this system were primarily formed by the weathering and removal of sediments from the Uncompahgre highlands to the east, and their westward transportation and deposition on the nearby lowlands by streams. The climate was undoubtedly hot and arid in Permian time, as it is now, and consequently the sediments were exposed to highly oxidizing conditions throughout the sedimentational processes, coloring the iron-rich sediments a rusty hue. The Permian strata today, as they are being exhumed from their final resting places by recent erosion of the high plateau country, are typified by beautiful red, reddish-brown, and purple shales that delight artist, photographer, and sensitive traveler. The red beds are exposed at the surface over wide expanses of the Colorado Plateau, earning the reputation for being the most scenic red desert country in the western world and the name "red rock country" for southeastern Utah in particular. Canyonlands lies at the heart of "red rock country," and the red fantasyland, is composed of this sequence of Permian red beds. Even if the Permian strata were not of scenic importance, they would be of extreme interest to the geologist for the tale they reveal when read from our natural geological history book of the earth.

The Permian Period of geologic time dawned on a setting that consisted of a broad lowland region that lay just above sea level throughout what is now Canyonlands and most of the Colorado Plateau. An exception was the magnificent Uncompahgre mountain chain that bordered the coastal plain on the northeast and east. Lazy, meandering streams wandered the low country, connecting the uplands with the shallow seas that lay a hundred miles or more to the west and south of the Ancestral Rockies region. The streams were heavily laden with sediments that had been ravished from the upland regions by the rampaging headwaters and tributaries of the now placid streams. Of course, the streams had a much greater gradient and carrying power while descending the mountain front, and consequently could carry large boulders as well as the finer grained sand and mud that the more sluggish stretches of river were able to move. As the Permian streams flowed from the steep mountain front to the gentler lowlands, much of their capacity to transport material was depleted, so the larger boulders and pebbles were dropped at the change in gradient, forming alluvial fans of sediments along the base of the mountains. The more sluggish rivers of the lowlands carried only the finer materials and distributed sand and mud over the lowlands in channels and floodplains before finally entering the western sea.

These processes were probably the same as those active in similar settings today, so the deposits of Permian time can be reliably interpreted by direct comparisons. Because of the complexities of studying the rocks in three dimensions and comparing them with two-dimensional modern models, the Permian rocks must be discussed not only by regions, but also by vertical layers in historical succession.

A thirty- to forty-mile-wide belt extending in a northwesterly direction parallel to the front of the ancient Uncompahgre uplift was the depositional site for the bulk of the coarse sediments being eroded from the highland. This mountain foreland extended southwestward from the main escarpment, which passed just east of Thompsons, Utah, and extended southeastward past Gateway, Ridgeway, and Pagosa Springs, Colorado. The gravel belt extended from that front across what is now the salt valleys of the Moab country, becoming generally finer in grain size away from the source area. The situation in this belt did not change very much from that of Pennsylvanian time, as sedimentation probably continued from the older period into the Permian and very thick accumulations of coarse arkose resulted. The 8,000 to 10,000 feet of Permian arkosic sediments cannot be subdivided into ordinary formations, as the conditions remained relatively constant for a considerable interval of time, and the red, coarse rocks are lumped into the "Cutler Formation" in that region. Fischer Towers and the adjacent canyon of Onion Creek, a few miles east of Moab, were carved from the upper few hundred feet of this undifferentiated Cutler Formation.

The salt-piercement structures were actively growing along this belt during Cutler time and strongly impeded the normal flow of sediments by continuously creating low hills as the salt domed the underlying rocks in the path of the Permian streams. The result was that the Cutler arkoses abutted the salt structures, or greatly thinned over them if, indeed, they could cover them at all. The streams flowed around the growing hilly topography and deposited the Cutler sand and gravel between and around the salt structures, creating extremely complex deposits of great thickness between the salt structures and little or no sediments across the tops of the growing features. A thickness map of the Cutler is very difficult to draw as a result of these extreme variations from several thousand feet of arkose to no arkose in a fraction of a mile when approaching a salt structure.

The growing salt walls further affected sedimentation by restricting sedimentation of the coarser grain sizes into very narrow bands paralleling the Uncompahgre uplift. By the time the Permian rivers found their way around the salt features, they were transporting only relatively fine-grained sand and mud, leaving behind most of the coarser debris within the region of salt structures. The addition of thick deposits in such a narrow belt caused continued growth of the salt wedges by creating a greatly localized overburden on top of the parent salt beds of the Paradox Formation. So while the salt-flowage features were rising and hindering the dispersal of Cutler sediments, the addition of the localized deposits of Cutler sediments was stimulating the growth of the salt structures. An impasse was created until the Uncompahgre wore down to nothing or the salt supply was depleted.

The second depositional belt, which also parallels the Uncompahgre and salt-valley trends, lies across the Canyonlands proper, and is composed of much finer-grained sediments of red and reddish-brown color. This finer-grained suite of rocks, or "facies" as geologists call it, extends westward from the salt- valley country for varying distances until stream deposits are seen to grade eventually into marine deposits. Depending upon the geographic position of the shoreline during various increments of Permian time, the exact rock facies vary rapidly, both laterally and vertically, across this depositional belt. The resulting complexities make it reasonable and desirable to recognize a number of individual formations within the lateral equivalents of the original Cutler Formation.

When this situation occurs, it is customary to call the entire section the "Cutler Group" instead of "Cutler Formation," since it contains the several new formations within the "group." So southwestward from the salt-valley country across the Canyonlands and adjacent regions, the Cutler Group is divided into numerous formations of distinctive lateral and vertical extent.

The lowest and oldest of these red-bed units is known as the lower Cutler, or Halgaito Formation, although it is now known to be of latest Pennsylvanian age. The red siltstones and shales are very extensive, reaching into the Monument Valley region along the Arizona-Utah border, where the name "Halgaito" originates, and on toward the southwest into the Grand Canyon. There the red beds comprise the lower member of the Esplanade Sandstone. This particular tongue of red beds does not appear to have been deposited in marine waters until well west of Grand Canyon proper, where the marine fossiliferous limestones of the Pakoon Formation replace the red beds in the section. West of Moab, however, the Halgaito interval bordered rather closely

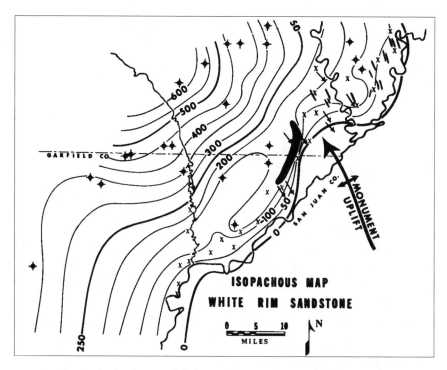

Map showing the distribution and thickness of the White Rim Sandstone in outcrops immediately west of the Colorado River along the western border of Canyonlands National Park and in wells drilled immediately west of the area of outcrop. The black sausage-shaped symbol represents a large offshore sand bar that grew in the White Rim sea, and the smaller bars toward the northeast are also shown in black. The arrows indicate the direction of the cross-stratification dip. (From Baars, 1962.)

on an extension of the sea near the present Green River and its confluence with the Colorado. Latest Pennsylvanian limestone, sandstone, and shale of the Elephant Canyon Formation with abundant marine fossils takes the place of the Halgaito red beds near the confluence of the two rivers, establishing a latest Pennsylvanian age and definite marine affinities for the section.

Brachiopods, crinoids, corals, and bryozoans are fairly abundant in this marine section, but another family of microfossils, known as fusulinid Foraminifera, are commonly present in the rocks and are most useful for dating and correlating the Pennsylvanian and Permian strata. The fusulinids were descendants of the endothyrid forams that were described as being useful in studying Mississippian sedimentary rocks, and are even better than their ancestors for the job at hand during the late Paleozoic periods. Although the fusulinids were still single-celled organisms, looking much like grains of wheat both in size and external shape, internally they were extremely complex, being rolled around an elongate axis of coiling, each volution being subdivided into numerous small chambers. Because of these complexities, the fusulinids must be sliced into thin sections, either along their axes or directly across their axes, for study and identification.

The dumping of coarse arkose into the inner gravel belt along the Uncompahgre upland continued uninterrupted into Permian time, when conditions were to change in the second facies belt. There, along a general northwest-to-southeast line that runs approximately through Grand View Point and Indian Creek, in Canyonlands National Park, the Cutler arkose abruptly met an incursion of quite distinctive white sandstone which entered the region from the northwest. The uninvited white sand was quite different from the arkosic Cutler sand, being relatively fine-grained and well sorted, and lacking the iron-rich red matrix of the local sediments. Furthermore, the white sandstone was invariably composed of thick beds of highly inclined depositional surfaces called cross-stratification. The inclined bedding indicates that sedimentation occurred on the steep leeward slope of either subaerial or subaqueous dunes; in other words, they were deposited either on sand dunes or on submarine bars, or both.

Regardless of their exact nature, the cross-stratification dips almost universally toward the southeast, indicating that the sands were transported from a northwesterly direction by either wind or marine currents. The main body of the white sandstone reaches thicknesses in excess of one thousand feet just west of Canyonlands National Park, and is extensive over the western reaches of the park, the San Rafael Swell to the northwest, the Monument Upwarp to the south, and the Grand Canyon region to the southwest.

It is most probable, after carefully studying the Cedar Mesa Sandstone, as the white strata are called, that the formation represents near-shore sand deposition along the eastern margin of an extensive sea that lay to the west during this late Lower Permian time. The cross-stratification probably was formed for the most part on small sand bars that lined the shallow underwater shelf area along the beach, but windblown dune deposits are locally present. The nature of the cross-bedding is more like water-deposited sand than dune sand, for the most part, and a few marine crinoid segments have been found within the formation. An abundance of glauconite, a distinctive green mineral formed only in seawater, and microscopic fossil foraminifera within the sandstone strengthens this interpretation. Longshore currents supplied the vast quantities of sand from somewhere toward the northwest, perhaps the general Salt Lake

City or the western Uinta Mountains region, but the exact source of the sand cannot as yet be located with any assurance.

The Cedar Mesa Sandstone and the arkose of the Cutler interfinger across a four- or five-mile-wide belt that extends through the Needles in southern Canyonlands National Park, and the Maze and Elaterite Basin, west of the Green and Colorado rivers. Although the magnificent facies change—from red arkose to the east and white sandstone to the west—can be seen from such vantage points as Grand View Point, it is best seen from a low-flying airplane, which gives an uninterrupted and godlike view of the transition. Long, thin beds of red sand may be seen to extend for several miles westward into the white rocks, and conversely the white tongues of Cedar Mesa Sandstone can be readily seen to invade the Cutler domain for a number of miles in an easterly direction. The overall change is gradual but definite, the white coloration becoming dominant toward the west as the red tongues gradually thin and eventually pinch out away from their Uncompahgre sourceland. The balances of sedimentation thus did battle across Canyonlands, and the alternating red and white bands in the Needles and Maze regions are testimony to transient and temporary victories as the battle between continental and marine sedimentation waxed and waned.

The belt of interfingering that marks the battlefront can be followed for many miles southward along Comb Ridge into the eastern Monument Valley country near Mexican Hat, Utah. The sandstone of the upper Supai Group, the Esplanade Sandstone, is an extension into the Grand Canyon country of the Cedar Mesa white sand facies, for the distinctive strata can be readily traced through deep wells into the Marble Canyon and Grand Canyon outcrops.

The Cutler red beds finally won the encounter, however, as a thick tongue of red siltstone and shale finally covered the Cedar Mesa Sandstone across the entire Canyonlands and Monument Upwarp regions. These deposits, known as the "Organ Rock Shale," comprise the red to brown slope-forming unit that overlies the Cedar Mesa over most of that region. It underlies the upper cliff-forming light-colored sandstone along White Rim in northwestern Canyonlands National Park and the cliff-forming sandstone that forms the vertical-walled monuments of Monument Valley. Thus, the Organ Rock Shale of the Cutler Group forms a distinctive red layer that separates two prominent cliff-forming sandstone units throughout much of the Colorado Plateau.

The Organ Rock Shale thins toward the west and northwest of the Green and Colorado rivers before reaching the vicinity of the present-day San Rafael Swell, but extends southwestward as far as Grand Canyon, where it is called the

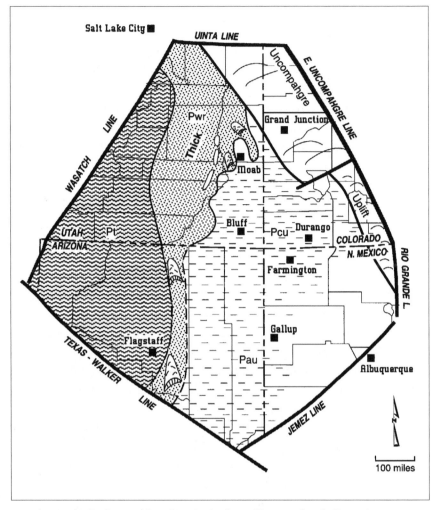

Map showing the distribution of the White Rim Sandstone (dot pattern) and adjacent time equivalent strata. Wavy pattern to the west indicates shallow marine deposits of the Toroweap Formation (Pt); Dash pattern is the equivalent Cutler red beds (Pcu).

Hermit Shale. Although these relationships are not altogether clear from viewing the outcropping strata in the several localities mentioned, the regions of thickening and thinning and pinch-outs is apparent when the findings of the several hundred deep wells drilled through the Permian System are taken into account.

The "white sandstone sea" made one last dying attempt to dominate the depositional regime in the Canyonlands region when it sent a final representative, the White Rim Sandstone, into battle against the red forces of the Cutler.

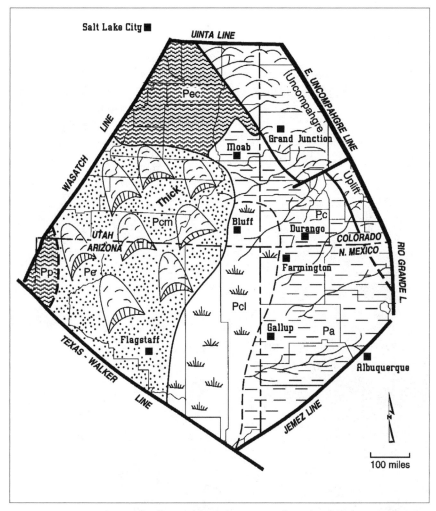

Map showing distribution of the Cedar Mesa Sandstone (Pcm) and its time-equivalent adjacent rocks. Counterclockwise from top: Pec = Elephant Canyon Formation; Pc = undifferentiated Cutler Formation; Pa = Abo Formation of New Mexico; Pcl = lagoonal facies of the Cedar Mesa; Pp = Pakoon Formation.

The formation was named because it holds up a broad, white topographic bench between slopes of red shale. It extends from just beneath Dead Horse Point, west of Moab, to Hite, Utah. The prominent marker bed thins gradually toward the east and pinches out approximately along the present-day course of the Colorado River. Because of this coincidence of pinch-out and recent erosional truncation, the White Rim Sandstone is exposed for many miles along the western borders of the Canyonlands west of the Colorado River, but is nowhere

View of the uppermost Permian and Mesozoic section along Shafer Trail, in Canyonlands. The cross-stratified sandstone forming the light-colored ledge in the foreground is the White Rim, overlying the soft-weathering Organ Rock Shale. Notice how the Organ Rock weathers readily along fractures to form pedestals. The middle shale slope is the combined Moenkopi and Chinle formations of Triassic age, capped by the cliff-forming Wingate Sandstone and the ledgy Kayenta Formation cap rock, both of Jurassic age. The dirt road in the lower right of the photograph will give an idea of scale.

found east of the Colorado. It thickens slowly but persistently from a feather-edge under Dead Horse Point to about one hundred feet thick where it dives beneath the Green River near Anderson's Bottom. The formation displays magnificent large-scale cross-stratification throughout that region. It has been shown conclusively that the White Rim interfingers to the west with the marine limestone/dolomite of the Toroweap Formation of Grand Canyon fame.

The White Rim Sandstone contains what appears to be remnants of ancient sand dunes near its eastern pinch-out, but includes large marine sand bars west of the Green River in and near Elaterite Basin. In the latter region, a fossil offshore bar measuring some 250 feet in thickness and ten miles in length has been uncovered by recent erosion. That it is a remnant bar and not a low hill on a pre-Triassic erosional surface is made obvious by the presence of well-developed, symmetrically shaped wave-ripple marks that cross over the top of the bar. Furthermore, the water-formed ripples are quite large along the flanks of the bar as a result of being formed in relatively deep waters, and become smaller in wave length as they approach the crest of the bar because of the shallowing of the Permian sea. Thus, the upper surface of the White Rim Sandstone in that region marks the sandy sea floor as it actually appeared in Middle Permian times, and outlines the presence of a large submarine sand bar that has since been buried and recently exhumed.

The offshore bar in the White Rim Sandstone in Elaterite Basin is presently filled with oil that seeps from the sandstone especially when it is made less viscous by the desert heat of midsummer. A number of tar seeps can be seen throughout the exposures of the fossil sand bar, and in places it is seen to run in sufficient quantity to fill a bucket in a few minutes' time. This phenomenon is fairly rare in nature and affords an opportunity to examine the nature of oil accumulations.

The oil in the White Rim Sandstone is not contained in large subterranean caverns, but instead fills the interstitial pore spaces between the individual sand grains. These tiny pores hold very little oil around any particular grain, but when all the untold billions of sand grains in a sand bar 250 feet thick by one mile wide by ten miles long are considered, a large enough volume of liquid black gold can be taken from the rock to make several oil companies wealthy.

Where did all that oil come from? It is difficult to say with certainty, but it is probable that the White Rim oil migrated up the eastern flank of the down-arched Henry basin, which lies immediately west of the exposures. There, the White Rim Sandstone is overlain by two petroliferous formations that do not crop out in Elaterite Basin. Both of these beds, the Permian Kaibab Formation and the Sinbad Member of the Triassic Moenkopi Formation, are dark-gray, petroliferous limestone and shale when cut by the drill, and either or both could be the "source beds" for the petroleum.

That oil rises to the top of water is a well-known fact. Since virtually all buried sedimentary rocks contain water, any oil that gets into a porous rock with included ground water will rise to the top of the bed because of the lesser density of the oil and the resulting greater buoyancy. If the strata are inclined because of some structural arching of the earth's crust, as in the case of the White Rim Sandstone, the oil will migrate up the inclined bed because of the same buoyancy relationships that make it rise in a bucket of water. Consequently, oil migrates up the dip of inclined, permeable strata until it is released at the surface or is trapped.

In the case of the Elaterite Basin sand bar, the natural pipeline of the formation was terminated abruptly up-dip when the east side of the bar was reached and the sand pinched out into impermeable red shale, creating a natural trap from which the oil could not escape. This kind of catchment mechanism is called a trap, and is necessary to localize sufficient amounts of oil to become economically significant.

Elaterite Basin sand bar was a dandy oil field, perhaps a couple of million years ago, before it was depleted by erosion. It now is only a skeleton of its former self that cannot be exploited by ordinary means. The more volatile gases have been released to the atmosphere and no driving mechanism remains to move the oil to a drill hole in quantity. The sand could, of course, be quarried, crushed, and treated with heat to drive off the oil, but that is too expensive for these times. And it would be far too expensive to remove the oil to market from this remote location. Anyway, the White Rim Sandstone is a fine example of a naturally exhumed and depleted oil field that can be studied directly at the earth's surface, without drilling wells.

The great Cutler red beds finally won the battle of the Canyonlands when a thin deposit of reddish-brown siltstone and shale finally buried the White Rim Sandstone in late Middle Permian time. These red beds are seen to drape across the Elaterite Basin sand bar, and occur at other localities along the Green River. The Permian sea retreated a bit farther westward, where it hesitated long enough to deposit the Kaibab Formation. Limestone and dolomite, similar to Kaibab exposures in the Grand Canyon region, extend northward along the western margin of the Colorado Plateau from Grand Canyon to the San Rafael Swell and beyond. Because it is uncertain that these exposures are truly Kaibab Formation or another similar unit, it has been called the Black Box Dolomite in the San Rafael country.

The supply of red arkosic sediments gradually dwindled and finally died as the Uncompahgre uplift, once a mighty mountain range, was lowered to a row of low, rolling hills by the relentless tearing away of the land by wind and rain. As the source of the arkosic debris diminished, the deposition of the Cutler Group ground to a halt, and a great era came to a close like an autumn sunset over "red rock country."

CHAPTER FOUR

BITS AND PIECES OF PALEOZOIC ROCKS

There are a number of smaller regions on the Colorado Plateau where land forms of spectacular scenic beauty have been carved from rocks of Paleozoic, primarily Pennsylvanian and Permian age. Because these several areas lie adjacent to the main Canyonlands region and their geologic history is closely related to that of Canyonlands, it is appropriate to discuss them here while the details of Canyonlands' geology are fresh in our minds. Since there is no particular advantage in proceeding in any particular order, the features of greatest interest will be discussed in geographic succession from northwest of Canyonlands southward and eastward around the margins of the Paradox basin.

The San Rafael Swell

A very large area of prominent up-arching of the layered rocks occurs to the northwest of Canyonlands. This presently uplifted structure, a magnificent example of an "anticline," is known as the San Rafael Swell. The structure lies southwest of the town of Green River, Utah, and northwest of Hanksville, Utah. As is usually the case with upfolded features, the higher parts of the fold have been most severely attacked by erosion. This is partly because the crests of folds are usually more highly fractured than the flanking strata, but also because the potential energy of running water is greater at high elevations and consequently its capacity to work is greater. Regardless of the reasons behind it, anticlinal folds are attacked more effectively by erosion near their crests, and the reverse is true of downfolded structures (synclines). Consequently, the axis of the San Rafael Swell has been deeply eroded and the older layers of sedimentary rocks have been exposed near the center of the uplift.

The oldest outcropping rocks on the Swell are believed to be of Mississippian

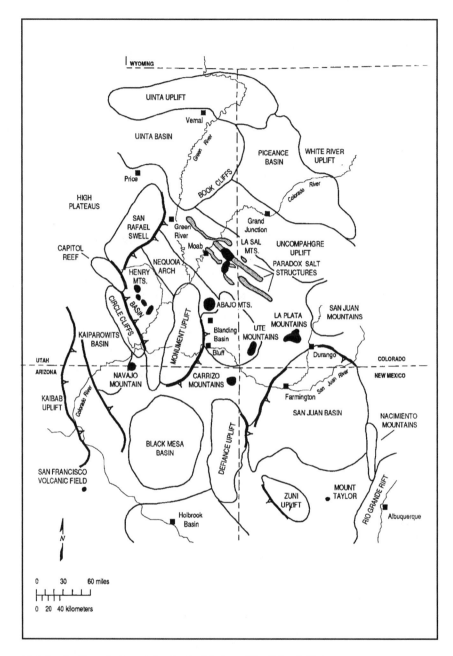

Map showing the major present-day structural features of the Colorado Plateau.

age, having been exposed by intensive erosion of a sharp canyon along the San Rafael River where it flows across the structural axis. The western flanks of the San Rafael Swell dip much less steeply than the eastern flanking beds, forming an "asymmetrical" anticline. Because of this, the structurally highest region lies east of the crest of the anticline. The axis extends for some sixty miles in a northeasterly direction, making the San Rafael Swell a very large structure indeed. The younger strata of Mesozoic age, which have been largely stripped from the crestal regions of the Swell, form prominent hogbacks. These flatirons effectively block access to the interior, forming what pioneer travelers called "reefs." The severe topography ("reefs") barred early wagon traffic, known then as "prairie schooners," onto the Swell for many tens of miles along the east flank of the uplift.

Pre-Pennsylvanian rocks are exposed only in one locality along the San Rafael River, but they have been penetrated repeatedly by wells in the probe for possible oil fields on the higher reaches of the anticline. The early Paleozoic section was found to be much like the previously described rocks in Grand Canyon and in the subsurface of Canyonlands. Here they become considerably thicker in a westerly direction, as the ancient Cordilleran seaway of western Utah and Nevada is approached. For example, the Cambrian System thickens from about 1400 to 2400 feet across the San Rafael Swell country, and becomes predominantly limestone and dolomite of marine origin similar to the Muav Limestone of Grand Canyon. No rocks of Ordovician or Silurian age have been recognized underlying the Swell, and the Devonian is mainly a thickening dolomite and sandy dolomite with no real sandstone present, such as the McCracken Sandstone to the southeast. The Mississippian strata are mainly thick, porous dolomites that contain no known oil fields such as those in the Canyonlands to the southeast.

One of the principal reasons why the San Rafael Swell has been barren of oil, even though large anticlines make excellent traps, is that the pre-Pennsylvanian rocks were exposed to weathering prior to Permian time. The Pennsylvanian rocks of the Paradox basin, which are quite thick a short distance to the east, have been completely stripped away from the San Rafael Swell. Thus, strata of Lower Permian age rest directly on Mississippian dolomites over and west of the axis of the structure. This unfortunate stripping of thick rocks was undoubtedly brought about by the uplift of the region in Late Pennsylvanian time, permitting the extensive removal of the Pennsylvanian strata and the weathering of the top of the Mississippian reservoir rocks. This early episode of uplift and erosion produced the so-called Emery uplift, which in effect was the ancestral San Rafael Swell of the Pennsylvanian Period. The relationships between this uplift and the larger Uncompahgre highland have not been completely ascertained, although the Emery uplift arose later and was buried earlier than the larger feature.

The Mississippian Redwall Formation is directly overlain by some five hundred feet of marine limestones, sandstones, and shales of the Lower Permian Elephant Canyon Formation. The formation thins westward, but is present throughout the San Rafael region, having covered the ancestral Emery uplift in Early Permian times. The Elephant Canyon strata appear to thicken toward the northwest into the Oquirrh basin in the Provo, Utah, region. Little is known of their extent once they pass beneath the thick Mesozoic and Tertiary section that comprises the High Plateaus country to the west. The Elephant Canyon Formation only crops out in Box Canyon of the San Rafael River on the east flank of the Swell.

A very thick, white sandstone, which appears to be much like the White Rim Sandstone of Canyonlands, overlies the Elephant Canyon limestone section throughout the San Rafael Swell. These highly cross-stratified rocks make prominent escarpments and slick-rock canyons along the crest of the anticline, and are now known to be the White Rim Sandstone. Early geologists were attracted to the geological exhibits of the San Rafael Swell and, seeing the massive, highly cross-stratified sandstone directly beneath limestone beds of the Kaibab(?) Formation, they quickly termed this unit Coconino Sandstone because of the general similarity of section between here and Grand Canyon. In those days, the Kaibab and Toroweap Formations were lumped into the single unit known as the Kaibab.

The exact relationship between the White Rim and the Coconino Sandstone have not been completely resolved to everyone's satisfaction. Studies of deep wells drilled between Grand Canyon and the San Rafael Swell now indicate that the White Rim Sandstone is equivalent in time and space to the eastern sandstone facies of the Toroweap Formation of Grand Canyon. The White Rim Sandstone was deposited at the same time as the Toroweap Formation of the eastern Grand Canyon region, and is now seen to be a shoreward equivalent of the marine Toroweap. It is clear, however, that the White Rim Sandstone of the Swell is not true Coconino. The classic Coconino is composed of windblown dune deposits with a southerly source that are known to pinch out just north of Grand Canyon. On the other hand, the White Rim Sandstone is marine or closely related to marine sedimentation with a completely different northerly source area.

The Totem Pole (right) and the Yei-bi-chai Dancers (left) in Monument Valley Navajo Tribal Park, Monument Valley, Arizona. The pillers and cliffs beyond are in the DeChelly Sandstone of Middle Permian age. The slopes below are the Organ Rock Shale, also of Mid-Permian age.

The limestone beds that directly overlie the White Rim Sandstone contain fossil brachiopods that reveal that they may be slightly younger than the true Kaibab. Because of the uncertainty, the "Kaibab" of the San Rafael Swell has been named the Black Box Dolomite. It is probable that the mid-Permian seaway lay mainly to the west of the Colorado Plateau country. Its eastern shoreline extended in a north-south direction along the eastern margins of the Kaibab and San Rafael uplifts; no remnants of the marine limestone have been found east of that line.

As is the case throughout the Colorado Plateau, "Kaibab" time marks the cessation of Permian sedimentation that was preserved on the San Rafael Swell. Red beds of the Mesozoic Era rest directly on the Middle Permian strata, with no indications that Late Permian rocks were ever deposited in the region. Some erosion occurred, which slightly sculptured the Middle Permian surface prior to Triassic time, but no evidence remains to suggest that younger Permian strata were deposited to any significant thickness. Perhaps the Late Permian was a time of low-lying lands on the Colorado Plateau that received little or no sediments, but was too low in relief to permit much erosion of the dominantly marine Permian section.

Monument Upwarp

Another uplifted region of major proportions extends southward from the Needles area in southern Canyonlands into northern Arizona, a total distance of about one hundred miles. Like the San Rafael Swell, this uplift has been excavated along its axis by recent erosion, exposing strata as old as Middle Pennsylvanian age in canyons of major drainages. This enormous anticline, called the Monument Upwarp, exposes several hundred feet of Pennsylvanian Honaker Trail and Paradox Formations in the deeper chasms of the San Juan River that crosses the crest of the uplift. The entire Permian interval of red beds at their grandest are exposed over vast regions of the upland desert. Actually, one of the deepest canyons, Cataract Canyon, which lies in the depths of Canyonlands National Park and has already been described, was excavated along the northwestern flank of the Monument Upwarp. And the intricate fracture and fault patterns of the Needles country lie on the northern plunge of the great uplift. The remainder of this discussion will be restricted to the southern half of the Monument Upwarp, where even-more-extensive exposures of the late Paleozoic section are to be found.

The San Juan River, a major tributary of the Colorado, flows across the width of the prominent Monument Upwarp with the same arrogance with which the Colorado flows across the Kaibab uplift through Grand Canyon. The San Juan was not as accomplished at dissecting the earth as the Colorado, but carved canyons of 1,500- to 2,000-foot depths for many miles as it crossed the uplifted structure. Erosional stripping was responsible for the denudation of the entire southern half of the Monument Upwarp. Here, the folding did not form a single, broad anticline, but instead it is a broad arched uplift composed of several smaller anticlines and synclines. These folds would ordinarily be considered major structures, forming textbook-quality displays of folded rocks that are obvious and self-explanatory to anyone. The canyons of the San Juan River cross several of these spectacular geological features, exposing them in cross section to the very core and forming a series of separate canyons as the river wanders from one anticline to the next.

The first of these canyons was carved through the Lime Ridge and Raplee anticlines, which constitute the eastern flank of the Monument Upwarp. There, the San Juan River first encounters the folds a short distance west of Bluff, Utah.

Monument Valley as seen from the air. This magnificent scenery sits astride the Monument upwarp, along the Arizona-Utah border. The prominent cliff-forming formation is the DeChelly Sandstone, of Middle Permian age, being gradually eroded back from the underlying soft slopes of the Organ Rock Shale. The buttes and mesas are capped by shaly beds of the Triassic Moenkopi Shale protected in their perched positions by a thin remnant of the Shinarump Conglomerate Member of the Chinle Formation. The low ridges in the background are the hogbacks of Permian and Mesozoic formations dipping steeply toward the east along the flank of the uplift. View is toward the northeast.

The river crosses upturned beds of the Mesozoic and Permian formations without difficulty, and then flows rapidly and majestically through spectacular gray gorges cut into the Honaker Trail and Paradox Formations that can be viewed on the surface only from a river boat. This trip is one of the best available, because in only one or two days of leisurely river running in rubber rafts, the very bowels of the Monument Upwarp can be seen in all their grandeur.

The first anticlinal axis encountered along the trip is the culmination of the Lime Ridge anticline, where rocks of the very topmost few feet of the Paradox Formation are exposed at river level. Close examination of the lower layers near the axis reveals that they contain the algal mounds that form the petroleum reservoir rocks in the south end of the Paradox basin. The mounded layers are composed almost exclusively of the algal genus *Ivanovia*, and the oily pore spaces that separate the fossil algae are abundant and large. The course of the river here leads an excursion through the truncated and naturally depleted oil field of the type that is so prolific in the Four Corners area a short distance to the east.

The Paradox beds again recede beneath river level as the shallow syncline between the two anticlines is crossed. When the axis of the second, or Raplee, anticline again brings the Paradox above river level, the algal bioherms have disappeared from the section, but here the deeper erosion has revealed the topmost several beds of gypsum and black shale in the Paradox Formation.

The entire Honaker Trail Formation is exposed above the Paradox beds, forming the characteristic ledgy gray canyon walls. From the fold axis the river rapidly crosses the west flank of the spectacular Raplee anticline, climbing up section through the Honaker Trail Formation in leaps and bounds until it emerges into the sharp Mexican Hat syncline near the desert village of Mexican Hat, Utah. There, the course of the river crosses the contact of the Honaker Trail and Halgaito Formations and proceeds to follow the axial regions of the syncline through the flaming red beds of latest Pennsylvanian age, past the rock monolith that looks like a Mexican peon sleeping with his sombrero on his head. The short river run ends here near Mexican Hat, where the river is accessible from the highway.

The more adventurous can continue the river excursion for several days through the next series of chasms cut by the San Juan River. The course of the river passes through the entrenched meanders known as the Goosenecks of the San Juan and the axis of the Mitten Butte anticline and then flows down the west flank of the Monument Upwarp. The Paleozoic strata plunge beneath the earth's surface into man-made Lake Powell that terminates the white-water adventure.

The scenery is much the same along this longer trip, but the river has cut the canyon into deeper geological horizons. Where the canyon plunges downward into strata where the Paradox evaporites should be, it is discovered that there are none in the section, and limestone beds have taken the place of the gypsum and salt—the southwestern margin of the Paradox evaporite basin has been crossed. In the vicinity of the Goosenecks, several hundred feet of limestone and shale of normal marine character are exposed beneath the Honaker Trail Formation, but no evaporities are to be seen. Just past the Goosenecks, near the deepest workings of the river, is Honaker Trail, the "type section" of the formation. The trail is an access route through the awesome gray cliffs of the San Juan Canyon, scratched from the cliffs for the exploration for placer gold by early prospectors. Limestone rich in fossils, even calcareous algae, are numerous throughout the length of the lower canyon, but the highly porous algal mounds are not to be found. Instead, numerous much smaller mounds are scattered through the lower half of the exposed Pennsylvanian section. They are here composed of fine-grained lime sediments and Foraminifera, and display only minor oil seeps.

Above and beyond the rims of the deeply entrenched river courses, the reds and reddish browns of the Permian Cutler formations create a fantasyland of colorful spires, buttes, and mesas. Here, the Permian strata are readily subdivided into alternating formations of soft, eroding shale and cliff-forming sandstone. The alternation of soft and hard beds creates stair-step topography and makes the carving of monuments and spires a relatively easy matter for the weathering goblins.

Monument Valley

Monument Valley, a Navajo Tribal Park, is perhaps the best known region of Permian rock sentinels, although several other spots on the Monument Upwarp are close rivals. The oldest of the red bed formations is the Halgaito Shale, now known to be of latest Pennsylvanian age, directly overlies the gray strata of the Pennsylvanian Honaker Trail Formation comprising the deep inner canyons. The contact between the older gray beds and the younger red beds is sharp and usually is marked by a thin bed of conglomerate, for it represents a relatively short break in the geologic record. The Halgaito red shale and siltstone beds were derived from the now distant Uncompahgre uplift when it was at its zenith. Only the finer-grained sediments were carried for such distant travels by the transporting streams, leaving behind the arkosic sandstone of more proximal regions of east central Utah and southwest Colorado. The Halgaito sediments were deposited on arid coastal lowlands that probably consisted of broad flood plains of sluggish rivers and intertidal mud flats and marshes along the seaway, which must have lain farther to the west in those days. These deposits are continuous with the lower Cutler arkoses in the Canyonlands region and the upper shales of the Supai Group in Grand Canyon. Halgaito strata are prominent in the lower slopes of Cedar Mesa just north of the Goosenecks of the San Juan and in the valley where the town of Mexican Hat is situated. It also forms interesting

West flank of the Raplee anticline from an aerial vantage point looking south. The strata exposed in the canyon of the San Juan River in the foreground are of the Pennsylvanian Paradox and Honaker Trail formations. The softer-weathering beds in the lowlands to the right are in the Pennsylvanian Halgaito Shale, which has been stripped back from the crest of the anticline by recent erosion. About 2000 feet of Pennsylvanian rocks are exposed in the cliff at the lower left.

topographic prominences such as Mexican Hat rock and other, similar buttes in this same general region. The formation also crops out in Cataract Canyon to the northwest, where it can be traced directly into the marine limestone and clastic rocks of the lower Elephant Canyon Formation.

The Monument Valley country was infested in Halgaito time with strange-looking four-legged creatures that looked something like large lizards. These awkward critters, that stalked the lowlands in search of food, were the ancestors of the dinosaurs. Bones of these primitive amphibians and reptiles have been exhumed in this region by erosion and paleontologists, indicating that

Aerial view of the Goosenecks of the San Juan, where the river crosses the crest of the Monument upwarp. The visible strata are entirely contained within the Pennsylvanian Hermosa Group, and nearly 2000 feet of section are exposed in the canyon. For scale, a two-lane paved road leading to an overlook point is visible in the upper left of the photograph.

the hot and probably arid climate of Halgaitoland was not inhospitable to such imbecilic and homely creatures as these.

The coastline of the Early Permian sea that lay to the west and northwest migrated into the central reaches of the Monument Upwarp, crowding back the Halgaito lowlands. A network of offshore and barrier bars not unlike the elongate sand deposits of our Gulf and southern Atlantic coastlines of today, formed along the western flank and axial regions of the uplift in the shallow coastal waters. Coastal dune fields formed at intervals, as the shoreline repeatedly advanced and retreated with the fluctuating sea level of the time.

The lower San Juan River canyon at the mouth of Slickhorn Gulch.
The lower half of the cliffs are the Honaker Trail Formation (Middle to Late Pennsylvanian)
overlain by slopes of the Halgaito Shale (Late Pennsylvanian).
The upper cliffs are in the Lower Permian Cedar Mesa Sandstone.

THE COLORADO PLATEAU

The resulting thick sandstone is today known as the Cedar Mesa Sandstone, named for its prominent exposures in the upper cliffs of Cedar Mesa, immediately north of the Goosenecks. There, in excellent exposures along cliffs and roadcuts bordering Utah State Highway 261, the cross-bedding that suggests the marginal marine origin of the formation can be clearly studied. The highly cross-bedded sandstones are here only a few hundred feet thick, but they become far thicker and more impressive toward the northwest until they reach a maximum of about 1,200 feet in the vicinity of Dark Canyon and Lower Cataract Canyon. Broad areas of the central Monument Upwarp, Cedar Mesa, are underlain by this formation, forming a high pinon- and juniper-shrouded plateau of formidable proportions.

It is into this upland that the sharp gorges of White Canyon and its tributaries have dissected the relatively comfortable plateau, forming a labyrinth of precipices, dry washes, intervening mesas, and natural bridges. The latter features are among the best of their type and have consequently been set aside, not surprisingly, as Natural Bridges National Monument. The bridges were carved from the Cedar Mesa Sandstone when the stream changed its course by cutting through narrow meandering "goosenecks" of rock. The undercutting process finally wore through the narrow extensions of rock around which the stream flowed, but left a natural bridge where horizontal beds of sandstone were strong enough to resist collapse into the newly formed channel through the rock. Such bridges actually span stream channels, in contrast to "arches" which form without the aid of stream erosion and stand alone.

A spectacular change in aspect of the Cedar Mesa Sandstone from nearshore sandstone to lagoonal red beds can be seen where the outcrops extend from Cedar Mesa eastward across the axes of the Raplee and Lime Ridge anticlines. In a matter of three or four miles the western sandstone facies gradually changes, by the addition of more and more red shale, into the eastern fine-grained facies along U.S. Highway 163 as it crosses the flexures. The Cedar Mesa strata may be traced into Comb Wash on the east flank of Lime Ridge anticline, it is seen to be a bright red shale containing gypsum nodules and stringers between the more brownish beds of the overlying and underlying formations.

This change of facies resulted from the same shoreline processes that formed the interfingering of the Cedar Mesa white sandstone and the Cutler red arkose beds in Canyonlands. Here the eastern environment was one of highly evaporative lagoonal waters behind the Cedar Mesa barrier rather than arkose-laden stream deposits. The evaporitic lagoon extended from the eastern Monument Upwarp into the Four Corners region, where it can be traced through several hundred wells drilled in the quest for petroleum. The eastern Monument Upwarp and central Canyonlands mark the easternmost extension of the white sand brought down from the northwest, and the western limits of sediments derived from the Uncompahgre uplift in late Lower Permian times. The white sandstone extended into Marble and Grand Canyons region, however, as the upper Esplanade Sandstone of the Supai Group.

The Cedar Mesa shoreline again retreated toward the west, and red coastal lowlands again stole the scene on the Monument Upwarp. This time the red beds were to be called the Organ Rock Shale for prominent exposures west of Monument Valley. The formation can be traced northward into the Organ Rock Shale of Canyonlands and the Hermit Shale of Marble and Grand canyons to the southwest. The source of the fine-grained red beds was the Uncompahgre uplift, which by this time was shedding only finer sediments; the prominent mountain range was being worn down to a rolling upland.

The Organ Rock Shale is usually a slope-forming unit lying between two prominent cliff-forming sandstone formations. It is best known for its occurrence in Monument Valley, where it makes up the lower gentle slopes of the buttes and mesas. There the valley floor is composed of the top of the Cedar Mesa Sandstone, overlain by the red Organ Rock carpet.

The prominent vertical cliffs of the monuments are carved from the next-younger unit of Middle Permian age, called the De Chelly (pronounced dee-shay!) Sandstone. This formation is unlike the other thick Permian sandstones we have already discussed, for it is orange-red rather than white and is probably of windblown origin. The formation of fossil sand dunes is known to extend over a rather large part of northeastern Arizona, northwestern New Mexico, and southeastern Utah, reaching a maximum thickness in excess of 1,000 feet between Monument Valley and Canyon de Chelly, from which it derives its name. The cross-stratification in the formation is typical of windblown (dune) deposits, being composed of thick, wedge-shaped packages of steeply inclined cross strata whose inclination to the horizontal reaches more than 30 degrees. Such steeply inclined bedding in sets that attain as much as fifty feet in thickness are forming today on the steep leeward slopes of sand dunes. They are caused by the sand being deposited on the steep lee side of the dune where the deposits become oversteepened and collapse in an avalanche of sand. The slope of the bedding plane is thus very steep, being the angle at which the avalanching sands come to rest, or approximately 34 degrees to the horizontal. If this origin is correctly interpreted for the De Chelly Sandstone, the dip of the cross strata points in the direction in which the

THE COLORADO PLATEAU

wind was blowing in Middle Permian time—the sand was being transported from the north and northeast by the prevailing winds. If this be the case the De Chelly sand was derived from the vicinity of the Uncompahgre uplift and the adjacent deposits of stream sand and gravel that probably acted as the source of the red sand. The windblown sand filled in the large basin forming south of Canyonlands country and east of the Monument Upwarp, but did not make it onto the top of the higher Kaibab uplift to the far southwest. Although the De Chelly Sandstone is in the same position in the stack of formations as the White Rim Sandstone of Canyonlands, the two units of divergent origin are never seen in close proximity to one another and probably are not continuous. There is growing evidence from recent drilling data that the White Rim Sandstone is the younger of the two, and correlates with the higher Toroweap Formation of Grand Canyon.

The spectacular buttes in Monument Valley are further capped by thin remnants of strata of Triassic age known as the Moenkopi Formation and the Shinarump Member of the overlying Chinle Formation, in ascending order. These beds, which form the thin culminating layers at the tops of most of the buttes, will be described in detail in the chapters on Mesozoic rocks of this region.

Canyon de Chelly National Monument and the Defiance Uplift

Still another elongate uplifted flexure lies along the Arizona-New Mexico state line, extending southward from the Four Corners area to U.S. Highway 66 and Interstate 40. Like the Monument Upwarp, the Defiance uplift, as this arch is called, is a north-south topographic feature composed of northwest-trending geologic structures, extending well over a hundred miles. It differs significantly, however, in that the Defiance uplift was a relatively high structure throughout Paleozoic time, and consequently Permian red beds are seen to lie directly on Precambrian quartzite along the crest of the uplift. Pre-Permian strata are known to underlie the Permian red beds surrounding the Defiance uplift in deep exploration wells. These strata of Cambrian, Devonian, Mississippian, and Pennsylvanian age onlap the positive feature in a succession that indicates that the Defiance area was a lowland throughout a great span of geologic time. The rather gently uplifted island supplied little or no sediments to the surrounding seaways, but effectively acted as a deterrent to marine travel between adjacent basin regions.

The country to the west of the Defiance uplift is known as the Black Mesa basin and apparently contained marine waters throughout most, if not

---

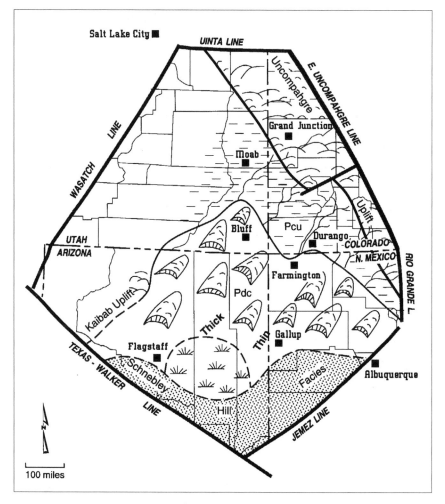

Map showing distribution of the De Chelly Sandstone (Pdc) and the adjacent Cutler red beds (Pcu). The dune pattern indicates areas of wind-blown (eolian) sand accumulation; swamp-grass pattern is area of evaporite deposits in the Holbrook basin; dot pattern in the south is an area of largely water-laid sand, known as the Schnebley Hill Formation.

all, of pre-Permian times. To the east, the region of the San Juan basin was an active seaway connecting the central New Mexico sea with the Paradox basin, during at least the Pennsylvanian Period, and perhaps much longer intervals of Paleozoic time. So its role as a successful barrier to communications between the basins makes the Defiance uplift of considerable geologic importance.

Canyon de Chelly is a gaping chasm carved from the northwestern flank of the Defiance uplift by the erosive processes of running stream waters. It is a magnificent flat-bottomed gorge whose vertical walls guard access to the bottomlands. The canyon was a natural habitat for the early Indian populations, whose ruined villages dot its inner realms. The red-walled canyon network has its beginning near the crest of the great uplift, and deepens until the dendritic system of tributaries converge into a single large drainage at a point where Spider Rock now stands guard near the steepening west flank of the structure. At that point, the vertical orange cliffs of the De Chelly Sandstone stand more than 800 feet above the exposed upper reaches of the underlying Cutler red beds.

A close look at the textures of the canyon walls, such as along White House Ruin Trail, which winds its way lazily through the treacherous sandstone walls in the western part of the canyon, reveals the multitude of bedding incongruities that comprise the formation. The massive vertical to overhanging cliffs of the De Chelly Sandstone display a family of interrelated individual units that lack regularity while creating intricate patterns of bedding architecture. If one segregates the constituents of the mosaic, it is revealed that the bedding is not horizontal, as is usually the case, but instead is composed of innumerable wedges containing steeply dipping cross-stratified layers that intermingle randomly. This is the basis for the interpretation that the De Chelly Sandstone was deposited as a Middle Permian dune field that developed along the coastal lowlands of the central Colorado Plateau region. Rare tracks of ancient amphibians trace the movements of primitive beasts that lived in the De Chelly wastelands. The canyon terminates where the sandstone formation dives beneath the surface toward the west, into the structural depression called the Black Mesa basin, and the sand-choked wash meanders onto the Arizona desert at Chinle, Arizona.

The De Chelly Sandstone can be traced onto the crestal regions of the Defiance uplift, where it thins appreciably to a minimum of only two hundred feet near Fort Defiance. This local thinning may reflect the still-high nature of the uplift. The De Chelly is not the first unit to cover the highland. Pre-De Chelly rocks have been exposed in a relatively small dry wash just northwest of Fort Defiance, where some six hundred feet of lower Cutler red beds are revealed. The red siltstone and earthy sandstone rests directly on the upturned and truncated quartzite of probable Precambrian affinities. No evidence remains that any of the older Paleozoic formations were present over the crest of the uplift.

The Defiance positive trend must have remained high throughout earlier Paleozoic time, only to be buried for all time by the Lower Permian clastic sediments from the Uncompahgre uplift to the northeast. Although buried in red debris, the Defiance structure may have remained somewhat high or been slightly rejuvenated in De Chelly time, for the thinning of the De Chelly Sandstone from eight hundred feet to only two hundred feet is localized along the axis of the Defiance arch.

South of Fort Defiance the sandstone cliffs appear to contain remnants of other formations that are better known in northern New Mexico. At Hunter's Point, about midway between U.S. Interstate Highway 40 and Fort Defiance, several thin units are noticeable in the formation which may include the Yeso and Glorieta pinch-outs. These formations are younger than the true De Chelly Sandstone and will be described in the next section. The

The De Chelly Sandstone as it appears along White House Ruin Trail in Canyon de Chelly National Monument. Here at its type section the De Chelly Sandstone is about 800 feet thick and is a fine-grained orange sandstone that displays the steep cross-stratification and irregular wedge-shaped packages of cross strata that typify windblown deposits. The trail in the lower right will give some idea of the scale.

THE COLORADO PLATEAU

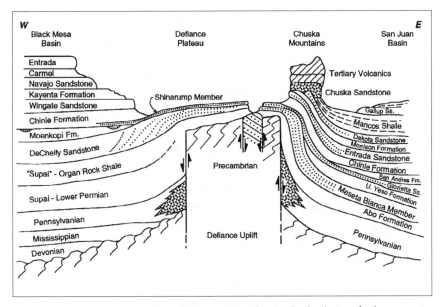

W

Black Mesa
Basin

Defiance
Plateau

Chuska
Mountains

San Juan
Basin

E

Entrada
Carmel
Navajo Sandstone
Kayenta Formation
Wingate Sandstone
Chinle Formation
Moenkopi Fm.
DeChelly Sandstone
"Supai" - Organ Rock Shale
Supai - Lower Permian
Pennsylvanian
Mississippian
Devonian

Shinarump Member

Precambrian

Defiance Uplift

Tertiary Volcanics
Chuska Sandstone
Gallup Ss.
Mancos Shale
Dakota Sandstone
Morrison Formation
Entrada Sandstone
Chinle Formation
San Andres Fm.
Glorietta Ss.
U. Yeso Formation
Meseta Blanca Member
Abo Formation
Pennsylvanian

Generalized cross section showing the distribution of rock strata across the Defiance uplift along the Arizona-New Mexico border. The uplift was obviously high during much of geologic time, causing numerous changes in rock characteristics between the Black Mesa and San Juan basins. Note the differences in names used in the two states.

entire Permian family of formations plunges southward under Mesozoic strata where the axis of the Defiance uplift crosses under Interstate Highway 40 just west of the New Mexico "Land of Enchantment."

Paleozoic Rocks of the Zuni Mountains

The Zuni Mountains provide a gently rolling, tree-covered oasis in the high deserts of the plateau country to the southeast of the Defiance uplift and just east of Gallup, New Mexico. The increase in elevation necessary for the climatic anomaly was provided by the presence of another major up-arched structural feature, known as the Zuni uplift. The axis of the northwesterly-trending anticline has been deeply eroded and exposed to the Precambrian core by erosion, revealing some similarities and many difference with its close cousin the Defiance uplift. The similarities include the fact that Permian red beds overlie the Precambrian basement on the uplift, but older marine strata are known to surround and onlap the structure in wells drilled along its flanks. This implies a very similar basic historical development for the two uplifts, for the Zuni must

---

Bits and Pieces of Paleozoic Rocks

also have been high during most or all of pre-Permian time. One of the more elementary differences include the presence of granitic rocks in the basement rather than the quartzite on parts of the Defiance.

A brief look at the Permian formations that overlie the granite on the Zuni uplift will provide a means of comparing the rocks of Cutler affinity to the north and northwest with the New Mexico formations. Here the names are different, but the formations have only slightly different connotations. The sediments were still derived largely from the Ancestral Rockies to the north, but the formations begin to mingle with beds of marine and evaporitic origins to the south and southeast of the Zuni region. Thus the Zuni uplift was a transitional environmental realm throughout parts of the Permian, and appears to have been a shoaling element during times of marine incursions.

Red beds of Early Permian age rest directly on the basement rocks over most of the Zuni Mountains exposures. Local remnants of very thin marine strata, the Bursum Formation, intervene near the southeastern plunge of the seventy-five-mile-long anticline. The arkosic red clastic beds are similar in physical aspect and age to the lower Cutler beds of the Defiance uplift and Monument Valley regions, but they are known as the Abo Formation in New Mexico. The fluviatile (stream-deposited) sandstone and shale beds thin to only two hundred feet as they bury the ancient uplift. That is less than half its normal thickness in surrounding wells. The formation attains a maximum thickness for northern New Mexico in the Albuquerque country to the east, where it surpasses eight hundred feet. Toward the southeast, the Abo interfingers with marine formations known as the Bursum and Hueco. These formations contain fossiliferous limestones of latest Pennsylvanian age, the Bursumian Series, thereby dating at least the lower Abo.

The next-overlying unit in the Zuni Mountains is an orange sandstone, not unlike the De Chelly Sandstone in general appearance, but here it is finer-grained and more massive. This distinctive unit is known locally as the Meseta Blanca Member of the Yeso Formation. It can be traced regionally into the De Chelly Sandstone through wells north of the Zuni uplift. It is probably water-laid in the Zuni region, for the typical wind-deposited sedimentary structures give way to massive and relatively flat-lying bedding as the unit thins southward to about 350 feet.

The upper member of the Yeso Formation, which might better be considered a separate formation because of its distinctive nature, consists of alternating thin beds of reddish-brown clastic rocks and two to four thin beds of dolomite throughout the Zuni Mountains. In addition to the more general rock types, it contains thin beds of gypsum in the easternmost exposures. Regionally, the evaporites thicken to become a prominent part of the unit, suggesting the highly evaporative climate at the time of sedimentation. Fossils from the Yeso Formation suggest that it was deposited in marine water. The presence of significant thicknesses of gypsum require that the marine basin of deposition be highly restricted as to the free influx of fresh, open marine waters. Gypsum forms by the direct precipitation of calcium sulphate crystals from sea water under conditions of intense evaporation, and salinity much greater than normal sea water. Consequently we would expect the Yeso rocks to represent sedimentation in a marginal marine basin whose circulation system was greatly hindered during times of intense evaporation and probably great aridity. Strata of the same age in West Texas are shallow shelf and open marine deposits, implying that the main seaway lay to the south during late-Lower to early-Middle Permian times. The Yeso Formation, at least in its upper member, is younger than the De Chelly Sandstone and is not represented in the Four Corners region.

A cliff-forming sandstone overlies the Yeso Formation in the Zuni Mountains that is quite distinctive from any of the light-colored sandstone formations yet described. It is in some respects similar to the Coconino Sandstone of the Grand Canyon country in that it is light tan to white and the grain size is rather fine as sandstones go. It differs, however, in the nature of its bedding. The Coconino, as you will recall, is composed of irregular wedge-shaped packets of cross-stratification whose individual beds dip very steeply because of their windblown origin. In the case of the Glorieta Sandstone of the Zuni Mountains, the cross-bedding occurs in relatively thin, horizontal units of only a few feet in thickness for the most part. The contained cross-bedding dips at only about half the angle of the Coconino on the average. It is most probable that this kind of sedimentary structure was formed in shallow, rather fast-moving currents of water, and that the cross strata represent sedimentation on the lee slopes of large underwater ripples or small sand bars. Because the Glorieta Sandstone is found to occur over vast areas of central New Mexico and because it is underlain and overlain by marine formations, it is most likely of shallow marine origin, even though it contains no known fossils. In a few isolated outcrops the formation includes large-scale cross-stratified sand bodies that appear to be offshore sand bars of considerable magnitude. However, these occurrences are quite unlike the windblown sand deposits of the Coconino, and should not be confused with them. The Glorieta Sandstone conforms to the propriety set by its neighbors, the underlying Yeso Formation

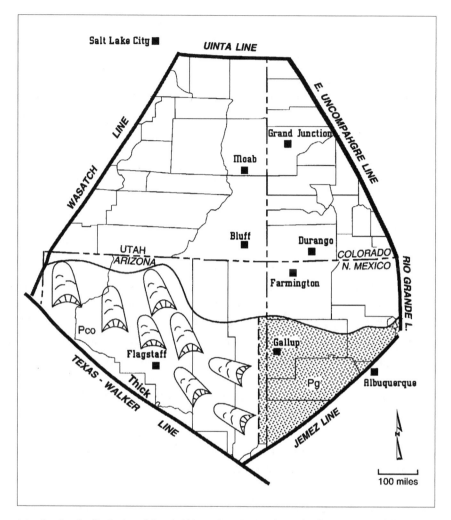

Map showing the distribution of the wind-blown Coconino Sandstone (Pco) in Arizona and the water-laid Glorieta Sandstone (Pg) in New Mexico. We can't see the exact "shoreline," but the change in character of the two sandstone bodies must lie along the axis of the Defiance uplift near the state line.

and the overlying San Andres Formation. It includes beds of gypsum and limestone to the southeast of the Zuni Mountains in central New Mexico. The Glorieta interfingers with the overlying formation in the northeastern part of the Zuni Mountains south of Grants, and disappears a short distance to the north of the ancient uplift in the subsurface of the San Juan basin.

The Glorieta extends westward from the Zunis and can be traced into the eastern extremities of the Coconino Sandstone along the Arizona border.

The two formations are obviously continuous and of the same geologic age. Details of the bedding of the two sandstone formations makes it apparent that the sand was derived from a land area in north central Arizona known as the Mazatzal uplift, in the vicinity of Globe. The prevailing winds blew the sands eastward toward the Arizona-New Mexico border, where the sand was contributed to the budget of the Glorieta sea. Thus, the similarities of the sand type between the two otherwise distinctive formations is justified, since they derived their sand from the same source area and the Coconino environment was the supplier.

The youngest Paleozoic strata in the Zuni Mountains are undoubtedly a continuation to the east of the Kaibab sea, which inundated the Grand Canyon region near mid-Permian time. The Kaibab Limestone can be traced southeastward and eastward from the Grand Canyon through outcrops that make up the surface of the flat plateau country throughout the greater Holbrook-Winslow-Flagstaff region toward the New Mexico border. Then the marine limestone beds are buried by younger strata along the state line between Holbrook and the Zuni Mountains. They are not present on the Defiance uplift a short distance to the north, possibly because the structure was rising slightly in Kaibab time and stood above sea level. The age-equivalent San Andres Limestone in the Zunis is similar in physical appearance and stratigraphic position to the Kaibab, and was most likely deposited along the northern limits of a much broader seaway that extended across most of Arizona and New Mexico.

The San Andres Limestone is extremely variable in its composition in the vicinity of the Zuni Mountains. It is, generally speaking, a formation composed of two distinctive units: The lower half is typically a thin-bedded mixture of dolomite, sandstone, siltstone, and shale that forms a ledgy slope between the cliffs of the underlying Glorieta Sandstone and the upper San Andres Limestone. The upper half of the formation is usually a more massive limestone that forms prominent vertical cliffs wherever it crops out. This more obvious member is most frequently found to be composed of lithified lime mud that occasionally contains large cephalopod and brachiopod fossils.

The type of limestone varies locally. For example, south of Grants, New Mexico, the formation is composed of lime sand that was probably deposited on a Permian beach. The ancient lime beach sand is intimately interbedded with quartzose beach sand of the underlying Glorieta Sandstone. Just north of the Zuni Mountain, the entire formation grades rapidly to red shale and siltstone, as seen in deep exploration wells.

Another very interesting locality for studying the San Andres Limestone

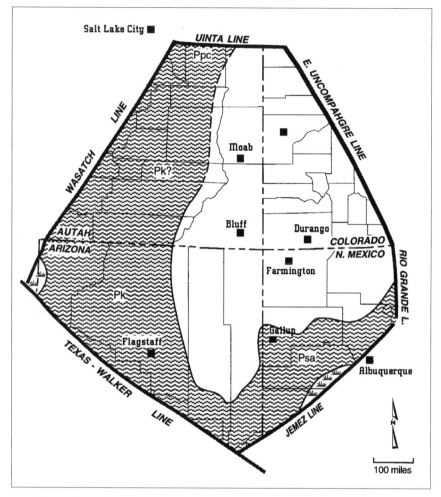

Map showing the areal extent of the Kaibab Limestone (Pk) in Arizona and the equivalent San Andres Formation (Psa) in New Mexico. In the northwestern part of the Colorado Plateau, the uppermost limestone may be somewhat younger than the true Kaibab, and probably should be called the Park City Formation (Ppc).

is midway between Fort Wingate and McGaffy along State Highway 400. In these northwesternmost exposures, the upper part of the San Andres is composed of a series of partially overlapping mounds of lime mud and fossil fragments forming bioherms. The mounds are thought to form in very shallow marine waters where organic life is extremely abundant and the various kinds of organisms proliferate freely and rapidly. When an individual organism, such as a brachiopod or an alga dies, its calcareous hard parts fall to the sea floor to become a part of the sediment accumulation. When the biologically produced lime sediments accumulate faster in one area than another because the living conditions are somehow better, a mound or bank is produced that is simply a "bone yard" of organic debris. Similar sediment mounds are forming today along the seaward shores of the upper Florida Keys, where marine grass and algae protect the shoals from marine erosion. Immediately to the north and northwest of the well-developed San Andres bioherms, post-San Andres and pre-Triassic erosion has removed all traces of the formation. Red beds of the Triassic Period have been deposited directly on the Glorieta Sandstone to form the northern limits of the San Andres Limestone in this region.

The San Andres Limestone in the Zuni Mountains appears to represent the deposits of shallow-water sediments on a broad, shallow shelf on the northern shore of the Middle Permian sea. The bioherms are certainly formed in very shallow water, perhaps only a few inches to a few feet deep, and of course the beach deposits mark the very shoreline. The river- and lake-deposited red clastic sediments of equivalent age that are found in the wells to the north of the Zunis surely suggest that continental conditions prevailed north of the uplift. Gypsum beds found in the San Andres Limestone toward the southeast, in the vicinity of Lucero Mesa, imply a deepening of marine conditions in that direction, but also a region of restricted water circulation and intense evaporation. Thus, the area now occupied by the Zuni Mountains appears to have been an anomalous shallow shelf in the San Andres sea, separating the evaporite basin to the south from the continental environments of the north. It is highly probable that the uplift along the Zuni axis was rejuvenated slightly in San Andres time to form these shoaling conditions which produced the variety of sedimentary environments found there.

This series of formations of Early to Middle Permian age are widespread in their occurrences throughout much of New Mexico. All the formations are to be found along either side of the Rio Grande trough that extends from the Albuquerque country southward to El Paso. In the Carlsbad country, however, the formations grade rapidly into complex, shallow marine shelf deposits that contain the well-known Carlsbad Caverns. Each formation has its marine limestone equivalent in southern New Mexico and West Texas, because the Permian seaway persisted across the region throughout the entire geologic period.

The only other scenic outcrops of the New Mexico Permian formations are found north of Albuquerque. In the vicinity of Jemez Springs, between the Nacimiento Mountains and the Jemez Mountains, the alternating red beds and light-colored marine beds of the Abo, Yeso, Glorieta, and San Andres formations provide beautiful scenery. Permian-age rocks occur for the most part above marine Pennsylvanian strata and beneath volcanic ash deposits of Tertiary age. The ash deposits fill a highly irregular topography carved into the Permian rocks prior to the volcanic catastrophe.

Permian sedimentary rocks younger than the San Andres-Kaibab formations are found in areas surrounding the Colorado Plateau, but no record of their existence has been uncovered here. Consequently, no formations representing the later half of Permian time are known on the Colorado Plateau, and the mid-Permian formations are blanketed by red beds of Mesozoic age throughout the region.

Glen Canyon Dam and Lake Powell, Arizona.
The dam blocks the flow of Colorado River through Glen Canyon, forming Lake Powell.
Suspension bridge (foreground) is 700 feet above the river.
Navajo Mountain is on the right horizon.

Monument Valley Tribal Park, Arizona/Utah, Navajo Nation.
Spring rains bring green to the high desert.
Monument names (from left to right, tallest group):
The Castle, The Rabbit, The Bear, and The Stagecoach.

Black Butte, north of Holbrook, Arizona.
One of numerous volcanic necks in the southern Black Mesa Basin.
Black Butte lies along the Pueblo Colorado Wash.
It is a classic example of a diatreme, an explosive eruption through sedimentary strata.

Hunters Point near Window Rock, Arizona, Navajo Nation.
Winter sunrise strikes fresh snowcover on dramatic folds of Hunters Point monocline
on the east flank of the Defiance Uplift.
Fog settles in the valley, shrouding Navajo homesites.

Near Tuba City, Arizona, Navajo Nation. Echo Cliffs (looking north),
a north-south trending cliff line marking the west edge of the Kaibeto Plateau (right).
Hamblin Wash and US Route 89 are in the valley below.

Marble Canyon, Glen Canyon National Recreation Area
& Grand Canyon National Park, Arizona.
The Echo Cliffs (left) tower over the upper reaches of Marble Canyon as the
Colorado River carves into Kaibab Limestone. Cathedral Wash enters from the right.

Kayenta, Arizona, Navajo Nation.
Agathla Peak (named El Capitan by Kit Carson) rises over 1,200 feet
above the desert floor south of Monument Valley.
The formation is the solidified neck of an explosive volcano, known as a diatreme.

Mexican Hat Rock, with community of Mexican Hat,
Utah in background and Monument Valley on horizon.
The rock itself is Cedar Mesa Sandstone resting gingerly on a bed of Halgito Shale.

CHAPTER FIVE

THE AMERICAN ALPS

THE SAN JUAN MOUNTAINS, COLORADO

From the depths of dark chasms carved into the earth's crust to the crests of magnificent alpine peaks we continue our journey through the history of the Colorado Plateau. The San Juan Mountains, our next destination, appear grotesquely in the plateau country of southwestern Colorado when the flat-lying strata of the organized world give way to the upturned ridges and crags of a province born by upheaval. The country, which geographically lies between Durango on the south and Ouray on the north, is a region which throughout most of its history has seen geologic unrest, culminating in a final up-arching or doming of gigantic proportions. The large circular fold has had its crest sculptured by erosion, as have all uplifts, and the eastern half of the range was masked by volcanic debris in Early to Middle Tertiary time. The uplands were finally excavated and partially whittled away by glaciers, that thrived during the great ice ages beginning only a million years ago. Glacial erosion exposed the older strata in grand style in the crest of the uplift and along its southwestern flank.

Because of the alpine setting of the higher parts of the range, some authors have included the San Juans with the southern Rocky Mountains province. The geology of the region so closely parallels that of the nearby plateau country, however, that it will be considered as part of the Colorado Plateau for the sake of this discussion. Thus, the San Juan uplift rises from the plateau like a gigantic boil, bringing up to the surface and exposing a segment of the larger Paradox basin along the flanks of the backbone of the continent; the Rocky Mountains of central Colorado.

The Grand Canyon has become a geologic legend in its own time because it exhibits, for close inspection, over four thousand feet of Paleozoic strata and a very spectacular canyon. However, the San Juan Mountains are not to be ignored in the study of the earth and its history. Sections of the earth's crust are exposed at Rico, Ouray, Durango, and elsewhere that display not only the Paleozoic section, but the Mesozoic as well. In the canyon carved into the uplift by the Animas River for about twenty miles immediately above and below Durango, a complete section is beautifully exposed that ranges from Precambrian granite up through an uninterrupted sequence of about sixteen thousand feet of sedimentary rocks of Paleozoic, Mesozoic, and Cenozoic age. The outstanding display of earth history culminates in deposits of glacial tills of Pleistocene age.

Although the Animas River canyon is not so deep as Grand Canyon, the mountains have a topographic relief of about nine thousand feet and are spectacular in their own right. The San Juan Mountains are, then, the antithesis of the Grand Canyon in that a greater section is exposed in a mountainous, rather than canyon setting. Both are magnificent.

Mountains of the First Era

The oldest rocks exposed in the San Juan Mountains are in some respects like the Vishnu Schist of Grand Canyon. The ancient metamorphic complex has undergone mountain-building revolutions. Even during the Precambrian Era it underwent such thorough alteration as to almost completely mask its original identity. The older Precambrian gneiss and schist, the Twilight Gneiss as it is called, consists mainly of dark-colored metamorphic rocks that may have originally been sedimentary rocks. In some localities a crude bedding can still be recognized. Beyond this guess, little can be said of its real beginnings. Whatever the nature of the original materials, they were subjected to the wrath of the mountain-building forces in very early geologic time. Metamorphism was completed by about 1.78 billion years ago in mid-Precambrian time. It had taken its toll prior to the deposition of the next sedimentary layers, which are still only of Middle to Late Precambrian age.

The drab-colored basement complex hosts some of the more rugged canyons that have been excavated from the domed mountains. The canyon of the Animas River, in its upper reaches below Silverton, was carved from the metamorphic sequence first by alpine glaciers and later to its present significant proportions by stream erosion. The canyon, although inspiring in its own right, also provides access to some of the most rugged alpine wilderness area in the United States.

During the summer months, the Animas River is tracked into the domain of the alpine demons by a narrow-gauge railroad, which provides a historically interesting and scenic visit to this fascinating region. The little train leaves Durango and slowly progresses up the lower Animas Valley through a complete geologic section of first Mesozoic and then Paleozoic formations that dip rather gently away from the mountain uplift. At first the route follows that of U.S. Highway 550 as it heads north toward Silverton, but at Rockwood the tracks leave the road to follow the mountain stream into the wilds. Immediately upon leaving the little village of Rockwood, the train wends its way into a magnificent canyon composed of the metamorphic complex and clings desperately to the wall of the abyss. The highly altered rocks of the first episode in recorded San Juan Mountain history can be seen in well-exposed outcrops for several miles along this stretch of the river. As the slow-moving train nears Needleton, in the heart of the mountainlands, the metamorphic terrain gives way to the pink cliffs of granite that were injected into the other Precambrian rocks some 1.5 billion years ago and is today known as the Eolus Granite. The granite can be seen to comprise the 14,000-foot-high Needles Mountains, to the east of Needleton. Almost immediately the tracks encounter a sequence of younger Precambrian quartzite and slate, but a fault that crosses the canyon at about Elk Park again brings the ancient basement rocks into view, and they are displayed magnificently until they plunge beneath the canyon floor just as Silverton comes into view. This is by far the easiest and most scenic way to see and study the older Precambrian metamorphic rocks.

The only place where the older rocks can be seen from the highway is where U.S. Highway 550 crosses Coal Bank Pass about midway between Durango and Silverton. There, for a mile or more immediately south of the pass, the gneiss abounds as the road traverses the flank of Potato Hill. The basement complex can also be seen with a little effort by climbing a short distance into the upper walls of Animas Canyon anywhere along the last five miles south of Silverton, where the road parallels the railroad some two thousand feet below.

The early Precambrian catastrophes that twisted the rocks into grotesque psychedelic textures were exhausted by Middle Precambrian time, and a new regime came into power. A series of sandstone and dark-colored shale was deposited in the Precambrian seas, where only a brief time before, there had been a mountainous region of untold magnitude. The clastic sediments attained a thickness of at least 10,000 to 12,000 feet, but the exact amount is anybody's guess because neither the top nor the bottom of the sequence is anywhere to be found. The water into which the sand and mud was dumped were probably of moderate depths, as shown by the very large ripple marks

The central San Juan Mountains of southwestern Colorado. The canyon of the Animas River, in the lower left of the aerial photograph, is cut into the ancient Precambrian gneiss and schist complex, which also forms the more rounded slopes in the center and right of the view. The jagged peaks on the skyline were carved by Pleistocene glaciers from a younger Precambrian granitic body called the Eolus Granite. These peaks are in the Needles Mountains, with three peaks at the right of the chain attaining heights of over 14,000 feet. They are reached best by way of the narrow-gauge railroad that occupies the bottom of the Animas Canyon.

(or "megaripples") that are often seen on the bedding planes of the sand layers. Sometime after the sediments were deposited, but before Paleozoic events began, the sedimentary layers were altered somewhat to form quartzite, which is highly cemented sandstone, and slate or metapelite, which is slightly metamorphosed shale. The degree of alteration was only slight when compared to the older basement complex, but the thick deposits of sedimentary rocks are in rather poor condition when compared to the younger Paleozoic strata.

The younger Precambrian sequence is called the Uncompahgre Formation for exposures of the highly folded strata in Uncompahgre Canyon immediately south of Ouray. The formation may be approximately the same age as the Grand Canyon Supergroup of red beds, which are sandwiched between the Vishnu Schist and Tapeats Sandstone in the depths of Grand Canyon. However, many geologists who have studied the two areas believe that the Uncompahgre is somewhat the older, as it is more highly metamorphosed than the Grand Canyon sequence. This may have resulted from a more severe history of tectonic activity (mountain building) in the San Juans, however, and may reflect nothing about the relative ages of the two units.

Again tectonic forces dominated the scene in the vicinity of the San Juan Mountains, for the Uncompahgre Quartzite and slate were down-faulted before the close of Precambrian time into elongate troughs of depressed rocks known as "grabens." The faults that caused all the trouble extended for many miles in a general northwest-to-southeast direction, but they are in a peculiar arcuate pattern around what is now the Needles Mountains. The magnitude of the displacement along the faults is not known, but it must have been on the order of several tens of thousands of feet in a vertical sense to accomplish the results now seen in the higher San Juan Mountains. The vertical movement along the faults is minor, however, as compared to the lateral displacement. The faults are now known to be "wrench faults," much like the San Andreas fault zone in Californian today, and lateral displacement may be on the order of tens, or even hundreds of miles. The belt of faulting has been shown to be of Continental scale, perhaps extending from the Bahama banks northwestward into Vancouver Island.

There are two of the curved grabens that dissected the region of exposures. One of these is exposed beautifully in Uncompahgre Canyon south of Ouray in the vicinity of the type section of the Uncompahgre Formation. The other fault block includes the country that now displays the Grenadier Range, across from Molas Lake south of Silverton. The bounding faults cross the highway at Molas and Coal Bank passes, respectively. In both down-faulted blocks the younger

Precambrian strata can be seen as steeply inclined beds that resulted from very tight and severe folds. In Uncompahgre Canyon the astute observer can see a tightly compressed anticline (upfold), especially if he has someone else to drive the car along the tortuous "million dollar highway."

In the Grenadier fault block the intense folding and related thrust faulting is not so easily recognized, for the view is usually from some distance. The upswept north faces of the nearly 14,000-foot-high crags along the Grenadier Range are the flanks of folds and thrust faults as displayed by exposed bedding planes in the quartzite. The intense folding and thrust faulting can be readily observed from the narrow-gauge train in the vicinity of Elk Park below Silverton. The intense structural distortions in the Uncompahgre Formation undoubtedly accompanied the large-scale down-faulting that placed the younger Precambrian rocks into juxtaposition with the basement complex, and accompanied continental-scale wrench faulting in the basement rocks that transected the deep Paradox basin.

The next major event was the intrusion of several separate bodies of molten rock into the two Precambrian rock types already described. The largest of these intrusive bodies is the Eolus Granite, which is found hosting the Needles Mountains. Several other granitic areas are known where the molten materials gobbled up the country rock and made their way into the preexisting rocks only to cool before reaching the earth's surface. This underground cooling process is slow and permits the growth of large crystals of the minerals that comprise the rock, forming "granitic" textures. The intrusives are known to be younger than both the other Precambrian rock types, for the gneiss and the Uncompahgre Formation have been partially replaced by the granite, a process that would be impossible if the two rock types were not already in place at the time the intrusion occurred.

The intrusive rocks have been dated by geologists of the U.S. Geological Survey, who determined the ratios of parent minerals to daughter minerals that form through radioactive decay. If the rate of this radioactive decay process is known and the relative amounts of original and secondary mineral matter can be determined, it is a simple matter to calculate the time involved since the formation of the parent mineral. The dates of cooling for several of the intrusive granites range between 1.4 and 1.8 billion years before the present. These ages of the granites fall well back into the Precambrian Era and verify approximate sequences of events that had previously been determined by the relative relationships of the rocks as mapped in the field. Since the intrusion of the granite was the last structural event in the history of the local Precambrian rocks, the obtained dates also tell us that both the older metamorphic rocks and the younger Uncompahgre strata were formed prior to 1.4 billion years ago: A very ancient date indeed.

Erosion attacked the region with a heretofore unknown ferocity, for the intrusive granites were exposed at the surface and the twelve thousand feet or more of Uncompahgre Formation was stripped from the flanks of the grabens, if it was ever present, prior to the time Paleozoic sedimentation began. These relationships are obvious because the Late Cambrian sandstone that initiated Paleozoic time in this region overlie or abut the granites and quartzites at various localities. It is common to read that the upper surface of the Precambrian deposits was planed off to a nearly uniform surface that was as smooth as a billiard table. However, this was not the case in the San Juan Mountains, as in Grand Canyon, where considerable topographic relief marks this great unconformity.

Unrest in Paleozoic Seas

The Paleozoic Era dawned, for all practical purposes, when the Cambrian sea finally arrived in southwestern Colorado. The shoreline had been advancing at a painfully slow pace across eastern Arizona and Utah after passing through the Grand Canyon country in about mid-Cambrian time. It did not arrive in the San Juan region until very latest Cambrian, or perhaps even earliest Ordovician, time. The dating of this significant event is not too accurate, because it is based entirely on the presence of tiny phosphatic brachiopod shells in the sediments that are difficult to date with any degree of confidence.

The sediments that were deposited in the advancing margins of the Cambrian sea in the San Juan country were not much different from the Tapeats Sandstone of the Grand Canyon, but no distinct fine-grained facies such as the Bright Angel Shale, or limestone like the Muav are found in the mountains. These rock types are restricted to the Paradox basin region, where they have been recorded in the cores and cuttings of deep wells. Despite the lithologic similarities, the Cambrian strata in the San Juans have been given a different name, largely because the correlation had not been made at the time the original geologic work was done in Colorado and the old name is thoroughly entrenched in the minds and literature of the region. The Ignacio Quartzite, as the local formation is called, was named for exposures in the vicinity of "Ignacio Reservoir," now known as Electra Lake, a few miles north of Durango, Colorado.

As already mentioned, the surface over which the Ignacio sediments were deposited was not worn smooth, but instead bore the traces of the earlier structural

---

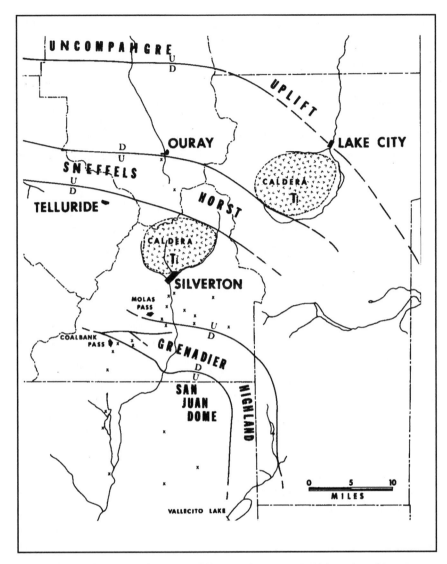

Map showing the distribution of the ancient fault system that was to control the geology of the San Juan Mountains throughout much of Precambrian and Paleozoic times. Notice the close relationship between the trends of the faults and that of the front of the Uncompahgre uplift. (From Baars, 1966.)

development. Paleotopography was in the form of peninsulas and islands of Precambrian rocks that jutted out of the Cambrian sea along the old faults. A good place to see these relationships is at Coal Bank Pass. There one can see exposures of Cambrian strata, that crop out for a few miles south of the pass.

One of the large faults that bounds the Precambrian graben crosses the highway a few yards north of the pass proper. On the north side, or down-thrown side, of the fault, the Precambrian rocks are Uncompahgre quartzite beds standing in a nearly vertical attitude, overlain by Devonian strata. On the south side of the fault and underlying the highway are the older meta-morphic gneissic rocks, overlain directly by the Ignacio Formation. Immediately adjacent to the fault, on the north side of the highway, the Ignacio is composed of a conglomerate whose boulders, derived from the Uncompahgre Quartzite, range upward to three feet in diameter. These boulders were de-posited as a talus on the flank of the ancient fault escarpment when the Precambrian beds stood in bold relief along the shoreline of the Ignacio sea. Across the highway to the south the Ignacio includes a basal conglomerate of much finer grain size and an upper series of sandstone. The change from boulders on the north to pebbles and sand on the south occurs within about two hundred yards. A mile farther south, where U.S. Highway 550 again crosses the Ignacio outcrop, the formation is seen to be a sandy shale and shaly sand-stone sequence, double the thickness as seen at the pass. The sea floor was sloping very rapidly into deeper-water conditions as the cliff along the old fault was left behind. Here, and at exposures that extend toward the south to Cas-cade Creek, the Ignacio thickens gradually and contains the tiny brachiopods with which the formation is dated, and the tracks and trails of trilobites, an-cient segmented arthropods that resemble a modern horseshoe crab.

In review, then, the Ignacio abuts the Precambrian fault at Coal Bank Pass, where it is a boulder conglomerate, and thickens rapidly while becoming much finer in grain size as one travels in a seaward direction relative to the Cambrian shoreline. The Cambrian sediments are not found immediately north of the Precambrian fault, for the area was above sea level in Late Cambrian time.

A similar series of rapid changes within the Ignacio Formation can be seen in the outcrops extending northward from Molas Lake to Silverton. The fault that provided the northern boundary of the Precambrian graben passes just to the southeast of Molas Lake, and boulder conglomerates of Late Cam-brian age again flank the fault. As one follows the outcrops northward, the conglomerate is seen to thin and disappear as finer-grained sandstone and eventually shale and thin beds of dolomite take over the formation. Again, the Precambrian fault block can be interpreted as a high escarpment along the Late Cambrian shoreline, the waters deepening rapidly in a seaward di-rection as the high peninsula of Precambrian rocks became more distant to the south. No Ignacio sediments were deposited on the fault block between

Coal Bank Pass and Molas Lake, for the Uncompahgre Quartzite is every-where overlain by Devonian or younger formations in that region.

The other Precambrian fault block occurs just south of Ouray. It appar-ently had a similar history, although it cannot be nearly so well documented. The Uncompahgre Formation, in the central portion of the fault block, is overlain directly by Devonian formations similar to the Molas—Coal Bank situation, but the margins are not exposed to show the rapid changes in the Cambrian sequence along its flanks.

Elsewhere in the San Juan Mountains, the Ignacio rests either on the older metamorphic complex or on Precambrian granites, having been deposited some distance away from the ancient structural elements in deeper waters.

It is obvious from these relationships that the Late Cambrian sea trans-gressed an area of southwestern Colorado where considerable topography on the Precambrian terrain existed. The topographic irregularities were largely controlled by the nature and geographic location of preexisting faults, which affected the basement rocks in pre-Ignacio times.

Rocks of Ordovician and Silurian age are missing from the San Juan Moun-tain outcrops, as elsewhere on the Colorado Plateau. No record exists of events that took place during some 100 or 150 million years of earth history.

The next recorded event to occur in the vicinity of the present San Juan Mountains was the readvance of the sea in Late Devonian time, following a long period of apparent quiescence. Sediments deposited in the marine trans-gression were largely dolomite of the Elbert Formation, but lenses and local layers of sandstone are common in the base of the formation and varicol-ored shale is often present throughout the unit. The sand probably was de-rived from the weathered surface of the then exposed Ignacio Quartzite, and represent near-shore and beach accumulations along the leading edge of the advancing sea. Most of the sand accumulations are concentrated along the flanks of the Precambrian grabens, suggesting that the offshore bars and other sand accumulations preferred the shoaling waters of the ancient structural features. The basal sand does not cross the ancient fault blocks, however, but in every case terminates at or very near the Precambrian faults. The obvious interpretation is that the early structures again produced topographic ridges in Elbert time which were not buried by the sand.

Because it is difficult to imagine a fault-formed escarpment remaining in relief for 150 million years, it seems more reasonable to assume that move-ment along the faults was rejuvenated in Devonian time. The remainder of the Elbert Formation, named for exposures along Elbert Creek between Coal

Bank Pass and Rockwood, is composed of dolomite and shale that appears to have been deposited on vast Devonian intertidal mud flats. The shaly to platy beds of the formation contain abundant evidences of stromatolites (algal mat laminations formed at the top of the tidal range in modern seas) and salt casts which undoubtedly formed just beneath the surface of the intertidal sediments under conditions of exposure in an intensely evaporating climate. These marginal marine mud flats almost buried the Precambrian fault blocks. The upper portions of the Elbert directly overlie Precambrian rocks across most of the high blocks.

Examples of this situation can be seen just north of Coal Bank Pass along lower Coal Creek, east of Molas Lake along lower Molas Creek, and south of Ouray at Canyon Creek and south along Uncompahgre Creek. All these areas lie within the ancient fault blocks. A fault block between Coal Bank and Molas passes was so high that the pre-Pennsylvanian formations all put together couldn't bury it. Consequently beds of Pennsylvanian age rest directly on Precambrian rocks on this small structural feature.

It should be noted at this point that the large Precambrian fault blocks have first been described as down-faulted blocks and later, in Cambrian and Devonian times, called high topographic features. While this probably seems contradictive, it is the observed situation in the field. These extensive basement faults are now known to be "wrench faults," where the dominant movement is lateral rather than vertical. In such cases, a yo-yo effect prevails, causing alternating vertical movement along the faults as seen in the San Juan Mountains.

Sea level gradually rose during the dying stages of Devonian time, for a limestone of marine affinities rests on the Elbert Formation with a gradational boundary. The younger formation, known as the Ouray Limestone, named for excellent exposures in the mouth of Canyon Creek on the outskirts of Ouray, occurs widely throughout the region of the Paradox basin. It is a dark-colored limestone that locally is altered to dolomite. Basically the Ouray is composed of a lime mud that was cemented into a limestone only after burrowing organisms exploited the sediments for their content of edible materials. Fossils are fairly common in the Ouray, and are mainly brachiopods, snails, crinoids, and forams, which suggest an age of latest Devonian to earliest Mississippian for the formation. Renewed movement along the now-buried fault blocks of Precambrian ancestry probably occurred in Ouray time, for the type of sediments changes markedly across the bounding faults. The Late Devonian movement of the main block seems to have again been down, for the Ouray strata are composed of dolomitized stromatolites (tidal-flat deposits) along the higher reaches of the flanks of the structure, and they abruptly change to limestone containing normal marine fossils within the faulted block proper.

Environmental conditions changed only very little as Mississippian time dawned in the San Juan region. Sedimentation was still restricted to lime sediments, but the resulting limestone and local dolomite are generally lighter in color and form more massive cliffs when exposed by weathering. The prominent ledges of the Early to lower-Middle Mississippian rocks are assigned to the Leadville Formation in the San Juan Mountains. The name was derived from the town of Leadville, Colorado, in the heart of the Rockies, where similar strata host rich metallic ore deposits.

The formation is part of a very extensive sheet of limestone that blankets the entire Rocky Mountain and Colorado Plateau regions, for it can be traced from province to province through outcrops and deep wells. It has been shown by several stratigraphers that the Leadville Formation of the San Juan country can be readily traced into the subsurface of the Paradox basin and Four Corners regions, and on into Grand Canyon, where it is known as the Redwall Limestone. Because of this obvious correlation of the two formations of similar rock type, fossil content, and age, we will use the term "Redwall Limestone" in this discussion to avoid confusion.

The Redwall Limestone is exposed in bold cliffs in the vicinity of Rockwood Quarry, some twenty miles north of Durango along U.S. Highway 550. There, the very pure limestone was quarried for its $CaCO_3$ content until recent years. Northward from the Rockwood area, it crops out along the route of the highway as far as Coal Bank Pass, forming a broad, gentle bench where the softer overlying Pennsylvanian formations have been eroded back from the apex of the San Juan dome. It again becomes obvious at Molas Lake, where it forms the bench upon which the lake resides, and northward almost to Silverton before it plunges beneath a thick volcanic cover. The formation does not emerge again until immediately south of Ouray, Colorado, where it can be seen forming abrupt escarpments along the west wall of Uncompahgre Canyon and lower Canyon Creek. The Redwall can be seen at other scattered localities, such as in Cunningham Gulch, east of Silverton, where it is the host rock to metallic ore minerals. These smaller exposures are sometimes difficult to locate.

The kinds of lime sediments that constitute the Redwall vary radically from place to place even though the general appearance of the formation changes only very little. In the Rockwood Quarry vicinity, for example, it is mainly made up of oolite and pellet lime sands that are cross-stratified to show that the site of deposition was one of sand bars. In the neighborhood

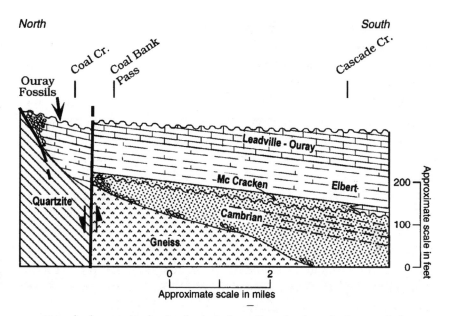

North   South

Generalized cross section showing changes in the pre-Pennsylvanian rocks along a major basement fault at Coal Bank Pass, north of Durango, Colorado. Note that the Cambrian Ignacio Formation thins, becomes coarser-grained, and pinches out against the fault. The Devonian McCracken Member of the Elbert Formation abuts the fault and is not present to the north. The upper shaly beds of the Elbert cross the fault, but lie directly on upturned Precambrian quartzite to the north. The Leadville-Ouray carbonate rocks cross the fault, but are intertidal deposits to the south and open marine with marine fossils to the north.

of Mill Creek and Cascade Creek on the south side of Coal Bank Pass, it is composed of dolomitized stromatolites, which are difficult to distinguish from the underlying Ouray Limestone. On the other hand, the Redwall contains crinoid-rich mud banks (or bioherms) at Molas Lake, Cunningham Gulch, and Ouray, that are almost identical to those in Canyonlands which produce important amounts of oil and gas, as already described. In sharp contrast, the entire formation was removed by pre-Pennsylvanian erosion from the crest of the Molas-Coal Bank ancient fault block, where Pennsylvanian strata overlie Ouray or older formations. These variations were probably caused by topographic relief on the Mississippian sea floor as the result of renewed movement along the Precambrian faults.

The crinoidal mud banks grew along the shallower shoals formed on the flanks of the fault blocks. Where the large faulted features became too high, so that they projected to or above sea level, the Redwall was completely removed by erosion. Similar relationships are known to occur deep beneath

---

the salt anticlines of the Moab region, where the crinoidal mud banks contain oil on the high flanks of the faults. The entire interval, including the potential reservoirs, was removed, however, by early erosion on more extreme faults such as the one underlying Paradox Valley.

We have now made a complete 360 degree swing in our discussion of the pre-Pennsylvanian stratigraphy and structure of the Paradox basin region. The ancient faults that were shown to underlie the salt anticlines of eastern Canyonlands rise out of the depths of the Paradox underworld to be exposed in the San Juan Mountains, where they can be carefully studied. The relationships associated with sedimentation of the Cambrian through Mississippian formations in close proximity to the ancient fault system can be seen to be almost identical in the San Juan Mountains and in the subsurface in the Moab country. The factors that were responsible for the development of oil-reservoir facies in the buried basin can be studied in exposures in the San Juans, and the knowledge gained can be utilized to cut exploration costs in the search for other accumulations.

Since these important faults parallel the edge of the Uncompahgre uplift of the Ancestral Rockies, they may be very closely related to the Pennsylvanian and Permian mountain ranges that provided the thousands of feet of red clastic debris to the Paradox basin. Were the Ancestral Rockies present in an incipient form in pre-Pennsylvanian and even Precambrian times? These are important points to be considered as more data are gathered to fill in the details of the historical development of this region.

The Rise and Fall of Pennsylvanian Seas

The curtains opened on the Pennsylvanian scene with the Redwall Limestone flooring vast lowland expanses. Warm, moist climates were actively producing a lateritic soil mantle across the entire Colorado Plateau and southern Rocky Mountains region. The San Juan country was not excluded from the action, for the Molas Formation is well developed as a classic fossil soil regolith in this region. The type section of the Molas is at Molas Lake, where the red weathering residue reaches some seventy feet in thickness. The formation does not represent a continuous record of soil production to such a thickness, but instead resulted from three related processes.

The exposure of the Redwall to the destructive processes of weathering resulted in the partial solution of the limestone and the subsequent concentration of the insoluble clay content into a residue that accumulated as a soil. The partially destroyed limestone can be seen in many roadcuts along U.S. Highway 550 between Rockwood Quarry and Coal Bank Pass, and at Molas Lake. The soil-forming process continued throughout Late Mississippian and Early Pennsylvanian times, but it was accompanied by a closely associated redistribution of the fine red soil material by streams that lazily coursed the coastal plains. The resulting sand- and gravel-filled channels and lenses are commonly present in the middle or upper layers of the Molas Formation in scattered localities throughout the San Juan Mountains.

Meanwhile, the Early Pennsylvanian sea slowly inundated the Colorado Plateau region: It came in from the south through the San Juan Basin from Albuquerque and down from the northwest (the Salt Lake country) into the northern Paradox basin. As the Pennsylvanian sea slowly overwhelmed the Four Corners region, the advancing wave front reworked the Molas sediments in its path, spreading them across the region in a thin veil of red, and mixing them with the remains of Pennsylvanian life. Fossil brachiopods and fusulinid Foraminifera can be collected from the top of the Molas at many localities in the San Juan Mountains, but one of the best collecting sites is in the type section at Molas Lake.

The Pennsylvanian sea reached the San Juan country somewhat later than in the Four Corners region proper, possibly because this was a slightly higher topographic feature resulting from the gently recurring uplift of the faulted dome. The marine limestone that typifies the base of the Hermosa Group in the Paradox basin, the Pinkerton Trail Formation, is only poorly developed in the San Juans. Relatively fine-grained sandstone and siltstone replaces the limestone.

The very early invasion of clastic debris is usual in the Paradox basin, and probably salutes the initial elevation of a southern extension of the Uncompahgre uplift known as the San Luis uplift. The San Luis positive region is now believed to be the fault blocks of Precambrian quartzites seen in the Uncompahgre Canyon and Grenadier Range areas of the high San Juan Mountains. It is not surprising that uplift of local segments of the Ancestral Rockies occurred at this time, for it is well known that the Front Range uplift—in approximately the same position as the present Front Ranges just west of Denver—began shedding coarse, arkosic sediments in Early Pennsylvanian time. The San Luis element must not have been especially high and rugged at this early time, because the sediments it shed were fairly fine-grained and not very thick. The resulting sand accumulations are thickest in the lower Animas Valley near the village of Hermosa, as if a large delta grew along the southern margin of the Paradox basin at this location. The sandstones thin rapidly and disappear from the section a short distance to the west and become rapidly less prominent up the Hermosa Cliffs toward the north. They thicken in northern New Mexico to form the Sandia Formation.

The accumulation of sandstone gave way up section in the Hermosa region to thick black shale and white gypsum layers of the southeasternmost extension of the Paradox evaporite basin. No salt is to be seen here, perhaps because the climate is too moist to allow salt to outcrop. Thin salt beds were drilled in a well just west of Durango. The only exposures of gypsum that are obvious to a viewer on the highway occur in Hermosa Mountain immediately north of the village of Hermosa, a few miles north of Durango. These evaporitic beds lie approximately midway up in the Pennsylvanian section, which is very well exposed in the vicinity. Early geologists in the region named the entire Pennsylvanian sequence the "Hermosa Formation" for these exposures in Hermosa Mountain and the Hermosa Cliffs, and for several decades the term was used throughout the Colorado Plateau for similar rocks of Pennsylvanian age. The term "hermosa" is Spanish for "beautiful," adequately describing these topographic features.

Later, in the mid-1950s. a great deal of refined study was leveled at the Pennsylvanian rocks in the Paradox basin because of their petroleum potential. The formation was elevated to group status and further subdivided into three formations—the lower Pinkerton Trail Limestone, the middle Paradox Formation, and the upper Honaker Trail Formation. Because the three newly defined formations were split from the parent Hermosa superunit and because they are still very closely related to one another, they are considered to be a part of a larger classification entity, the Hermosa Group. This is the usual route of stratigraphic nomenclature as more detailed studies are undertaken, for now each of the three new formations can be subdivided into smaller formal units called members.

The Paradox Formation of the Hermosa Group crops out in the type section at Hermosa Mountain, but is not seen again in the Hermosa Cliffs as they make their way northward toward their culmination under Engineer Mountain at Coal Bank Pass. This is partly due to the fact that the lower half of the Hermosa strata are usually covered by talus and other scree from the upper Hermosa Cliffs, but also the evaporites apparently thin rapidly and change to dolomite and shale to the north and east as the margin of the evaporite basin is crossed.

The upper prominent cliffs in Hermosa Mountain and the Hermosa Cliffs to the north are composed of alternating beds of sandstone, shale, and limestone that overlie the Paradox gypsum and black shale of this region. It is interesting to note that detailed studies of the stratigraphic relationships and the contained fossils tell us that the entire section exposed in the San Juan Mountains was deposited simultaneously with the thick section of salt that lies only a few miles to the west in the more central Paradox basin. This rapid change in rock type, from salt and gypsum in the basin to limestone and sandstone in the San Juan country, implies that the evaporite basin was already beginning to shrivel up in Middle Pennsylvanian time. This area was a shallow shelf bordering the true salt basin during deposition of the upper strata.

Almost any one of the Hermosa beds when studied in detail will demonstrate that the margin of the highly saline marine basin is near at hand, and gets closer as the beds are traced northward toward Silverton. For example, the beds of gypsum and black shale can be seen to grade northward along the Hermosa Cliffs first into massive beds of dolomite and then to thick limestone that contains abundant fossils, including calcareous algae that form the reservoir rocks in the Four Corners area. Finally, these limestone beds thin and become sandy, and change to sandstone layers near Silverton.

The Hermosa strata are well exposed along U.S. Highway 550 between Coal Bank and Molas passes, where the details of these changes can be studied. It is not uncommon to find the limestone to contain abundant marine fossils in this region, and even to construct fossiliferous mud banks (bioherms) along the highway and in beds visible from the highway. Any of these strata can be traced across Molas Pass, where they thin noticeably and change to beds of sandstone only a fraction of the original thickness as they approach the limits of outcrop just south of Silverton.

Another interesting complexity is apparent when the rocks in the Coal Bank Pass to Molas Lake region are examined with a wary eye. The Hermosa is there composed of a sequence of alternating limestone beds containing marine fossils and bioherms, interspersed with sandstones whose cross-stratification and other sedimentary structures were formed by stream deposition. In other words, the formation is composed of alternating beds of marine and continental origin. How can this happen?

It has already been mentioned that sedimentation in the Pennsylvanian seas, both in the Paradox basin and elsewhere in the world, was cyclic in nature. In the Midcontinent and eastern parts of the United States, Pennsylvanian strata are composed of alternating marine and continental layers, where limestone marks the marine phase and coal beds mark the continental phase of the cycle. Dozens of these cycles exist in some regions of the east, where they are called "cyclothems." Detailed studies in the heart of the Paradox basin have revealed the presence of cyclic deposits in that region also, but there the cycles are almost entirely marine and the layers were rarely exposed to permit continental sedimentation to occur between marine limestone and evaporite cycles. In the case of the central San Juan Mountains, where the margin of the great marine evaporite basin was close at hand at all times, slight fluctuations in sea level made considerable differences in the position of the shoreline. When the sea would rise a few feet onto the broad, flat, bordering lowlands, the position of the shore might migrate eastward for several miles, inundating thousands of square miles of previously useful land areas. In these shallow marine environments, lime sediments rich in faunal debris would accumulate rapidly. Then, in a few thousand or a couple of million years, the sea would again lower a few inches or a few feet and the shoreline would limp back toward the basin center and expose vast expanses of former sea bottom to continental processes. Thus, the shoreline waxed and waned across the San Juan region in Middle Pennsylvanian time, first depositing lime sediments, then clastics in stream channels.

Why would such fluctuations in sea level occur, especially on an area the size of a continent? The answer to that has been eluding geologists since the first cyclothems were studied decades ago. One hypothesis has it that throughout the continent, repeated uplift and subsequent sagging of the earth's crust due to mountain-building forces occurred, alternately lifting and then depressing the crust in a rhythmic manner. "The old yo-yo in the earth trick." Another proposed explanation is that extensive glaciation occurred, alternately taking up large quantities of sea water to make glacial ice, and releasing it when the glaciers melted during climatic cycles of warm and cold. This was known to alter sea level by perhaps several hundred feet during the just-ended "great ice ages" of the past million years. This seems to be a plausible explanation in the case of the Pennsylvanian and Early Permian sea-level fluctuations, for extensive deposits of glacial till are known to occur in rocks of Pennsylvanian and Permian age in the earth's polar regions, and fluctuating sea levels have been noted on a world-wide basis during these times. Since the glaciation is known to have occurred on a grand scale and sea-level fluctuations were world-wide rather than continental in extent, it seems likely that the glacial theory is the more plausible of the two. Consequently, the rise and fall of the seas in the Paradox basin was caused by variations in climatic cycles in other parts of the world, but which were capable of altering the nature of sedimentation significantly in this region.

Permian Floodplains Revisited

Permian strata of the San Juan Mountains are similar in many respects to the upper Paleozoic red beds of the eastern Paradox basin, although they do not reach such formidable thicknesses. These, the oldest red beds in the area, are called the Cutler Formation because of the excellent exposures in the vicinity of Cutler Creek just north of Ouray. There the formation attains a thickness of about two thousand feet and is mainly composed of stream-deposited arkosic sand and gravel with interbedded red mudstone and silt-stone. The sedimentary structures that typify sediments deposited in a flu-viatile environment (cut-and-fill structures, small-scale cross-stratification, lenticular bedding, and so forth) are well displayed in exposures along U.S. Highway 550 for a few miles north of Ouray and in Cutler Creek canyon itself. The high content of feldspar grains in the sands reveals that the source of the sediments was a nearby rock of granitic composition, and the pebbles and boulders present in many beds betray a relatively high source area not far distant from the site of deposition.

These characteristics result from the close proximity of the Uncompahgre uplift source area, which probably rose from the plains only a few miles to the east and northeast of Ouray during Pennsylvanian and Permian time. The Cutler Formation thins very rapidly toward the east in Cutler Creek and in the north wall of the Amphitheater, just east of Ouray. This thinning is due to the presence of a well-developed angular unconformity at the base of the Triassic red beds, formed when the Cutler was tilted and partially eroded away before deposition of the Triassic began. The unconformity completely truncates the Cutler toward the east in the Amphitheater above Ouray, per-mitting the Triassic to rest directly on Pennsylvanian strata to the east along the northern basement fault block.

The Cutler Formation is also well exposed in the canyon walls of the Animas Valley north of Durango. There the formation is not so coarse-grained, because it is somewhat farther removed from the source of the clastic debris, but it attains a thickness of twenty-five hundred feet of alternating red shale, siltstone, and pink, arkosic sandstone. Even a few thin beds of very dense limestone are present scattered throughout the red-bed sequence, suggesting that ephemeral lakes were common on the Cutler lowlands. No angular dis-cordance of bedding dips exists at the top of the formation in this region, but an erosional surface replete with deep stream channels marks the bound-ary between the Permian and the Triassic formations.

The Cutler is well displayed at many other localities in the San Juan Mountains, such as in the Piedra River Canyon, at Rico, at Telluride, and elsewhere, and invariably it overlies the ledgy gray cliffs of the Hermosa Group.

Summary

The Paleozoic and Precambrian rocks of the San Juan Mountains constitute a prominent part of the scenery and rugged splendor of the "American Alps." They are far more significant than that, however, for they represent an uplifted and exposed segment of the southern Paradox basin. Careful, detailed studies of these sedimentary rocks and their intimately related structures, the ancient fault system, have enabled geologists to interpret the deeply buried segments of the ancient evaporite basin with more confidence in their quest for petroleum products. Here, where the outcropping rocks can be thoroughly examined and collected, is where the models are born that are the primary tools in unraveling the mysteries of the buried and inaccessible depths. Application of principles learned from studying the exposed rocks to the deeply buried problems in the basin is still the best and cheapest means of exploration, even in these days of modern scientific technology and all the advantages it provides us.

PART TWO
Post-Paleozoic Landscapes

Elaterite Butte near the western border of Canyonlands National Park, Utah. The broad bench in the foreground is on the Permian White Rim Sandstone. Slopes at the base of the butte are the Triassic Moenkopi and Chinle formations, capped by the cliff-forming Wingate Sandstone and a thin layer of Kayenta Formation, both of Jurassic age.

CHAPTER SIX

FORESTS AND URANIUM

TRIASSIC TIME

Millions of years passed after the withdrawal of the Permian seas toward the west and south, but the marine waters did not return to the Colorado Plateau. No record remains to elucidate the events of the Late Permian, for rocks of Triassic age directly overlie the Lower to Middle Permian Cutler or Kaibab formations with only an erosional unconformity to mark the passing years. Triassic time was a period of relatively high continental areas and vast deposits of red sediments throughout much of the earth's surface. And the Colorado Plateau was no exception. The sea remained in seclusion to the south and west in the Cordilleran seaway, and deposits of marine origin can be found only along the western margin of the Colorado Plateau and in western Utah.

On the Colorado Plateau, however, continental conditions predominated throughout the geologic period, and great thicknesses of red strata resulted, adding to the color of broad regions of the province. The sediments were still largely derived from the Ancestral Rockies to the northeast, which were standing in low relief, and were strewn about the Colorado Plateau country by streams. The result is that the bulk of the Triassic sediments are fluviatile in origin, having been largely deposited in stream channels and adjoining floodplains along broad coastal lowlands, with marine strata becoming more prominent to the west. The plateau country, then, acted as a broad lowland region in Triassic time, with little or no effects of the individual Paleozoic basins and highlands interrupting the distribution of the sediments.

The principal cause for the present-day distribution of these early Mesozoic strata was the way in which the major uplifts and depressions that formed in Late Cretaceous and Early Tertiary time were affected by Recent erosion.

As already discussed, the axial regions of uplifted folds are always attacked first and most viciously by erosive forces. The crests of these uplifts are invariably more deeply carved and exposed to view than their downfolded cousins, the basins. This results in the younger strata being removed from the anticlinal crests, exposing the older rocks. The rocks of Triassic age are found mainly in broad bands surrounding the major uplifts, such as the Kaibab, Capitol Reef, San Rafael, Monument, Defiance, and Zuni uplifts, for the relatively soft sandstones and shales of the Mesozoic periods have been removed from these high folds by wholesale erosion. They have generally been preserved in the downfolded basins, but there the Triassic formations are still concealed by younger rocks of one kind or another. Consequently, the Triassic red beds are usually found in geographic belts that circumnavigate the uplifts.

Lower Triassic Mud Flats

The Mesozoic curtains open on a stage dominated by extensive brown tidal flats bordering seascapes that lay along the western margin of the plateau country. The resulting brown mudstone is known as the Moenkopi Formation, named for an Indian settlement a short distance east of Grand Canyon near Tuba City, Arizona. The sediments of the Moenkopi are almost universally composed of mud and silt, probably brought down from the very low Uncompahgre upland by streams and distributed by tidal currents traversing the mud flats in tidal channels.

The formation, generally speaking, changes most emphatically from east to west in a definite pattern. In the eastern Colorado Plateau of southwestern Colorado and northwestern New Mexico, the Moenkopi is either missing or is thin and composed of deposits that suggest largely a fluviatile origin. As one progresses westward, the formation gradually thickens and contains increasing amounts of mudstone and shale, which are horizontally bedded with extensive development of symmetrical and current ripple marks, mud cracks, and other surface features of the bedding planes that require shallow tidal-flat conditions to form.

These obviously shallow-water clastic sediments are best displayed in exposures along the western border of Canyonlands National Park and along the Little Colorado River drainage between Winslow and Cameron, Arizona, where the Triassic outcrops skirt the eastern Kaibab uplift. Similar rock types occur surrounding the Monument Upwarp, the San Rafael Swell, Capitol

Reef National Park, and elsewhere throughout a broad belt extending in a north-south direction across the central Colorado Plateau. In this region, the Moenkopi Formation crops out as a brown, slope-forming unit between more-resistant formations above and below, making a broad bench, notch, or valley wherever it is found.

West of the intertidal mud-flat facies the Moenkopi gradually displays more and more marine characteristics as it thickens to a maximum of two thousand feet. Marine limestone beds become prominent until they form a major part of the formation along the western margin of the Colorado Plateau, especially in the vicinity of St. George and Zion National Park in southwestern Utah. Still farther west and northwest, the Moenkopi grades into a totally marine section known as the Thaynes Limestone, which contains ammonitic cephalopods of Lower Triassic age. The marine phase can be seen in the foothills of the Wasatch Range just east of Salt Lake City, where it forms prominent gray ledges of limestone.

Tree Trunks, Stream Channels, and Uranium

Strictly continental environments dominated the scene at the close of Moenkopi time. A multitude of rivers and streams coursed the Colorado Plateau from border to border, meandering randomly over the surface to form an intricate interwoven network of stream deposits of wide geographic extent. The resulting deposits form the basal members of the Chinle Formation. In the southern half of the Plateau in northern Arizona and as far north in Utah as White Canyon near Natural Bridges National Monument, the basal unit is known as the Shinarump Conglomerate Member. It is generally a coarse sandstone and conglomerate deposited in complexly interlaced channels that cut into the Moenkopi red beds below. The stream deposits form light-colored cliffs between the soft, varicolored shales above and below where it is well developed, but it is often missing or thin in regions that the Middle Triassic streams did not reach. The name comes from the gray, hummocky Shinarump cliffs, which reportedly were named for their physical appearance; "shinar" meaning "wolf" in the Paiute Indian language and "rump" meaning a posterior in English.

North of White Canyon, another, somewhat younger series of stream sands and gravels takes over the role of the basal resistant member of the Chinle Formation and is known as the Moss Back Member. In the vicinity of White Canyon, the Moss Back is seen to overlie the Shinarump with an

---

intervening shale acting as a separator. The Moss Back is in most respects like the Shinarump, and indeed it was considered to be the same unit before the overlapping relationships were discovered.

Fluvial sediments are deposited in a number of characteristic but differing patterns directly related to the processes that formed them. In most regions where stream deposits are being formed, the gradient of the stream is low and the flow is retarded accordingly. The result is that the stream channel usually wanders about the broad floodplain in a rather erratic manner, simply following the path of least resistance wherever it might lead. The largest sand grains and pebbles are usually contained within the parts of the channel where the flowing velocities are greatest, and they are often deposited in the channel proper when the velocities are reduced for one reason or another.

The bulk of the waters move in definite parts of the channel depending upon the channel's shape as it meanders towards the sea. The strongest flow is directed toward the outside of the turns, where the water meets the bank with the greatest force, and consequently this bank is subject to erosion rather than deposition. The weakest flow in a meandering stream pattern is along the inside of the turn, as the main current is directed at the opposite bank, permitting suspended and rolling sediments to settle to the bottom and be preserved. The latter sediment accumulations are crescent-shaped deposits reflecting the shape of the inner side of the curve, and are called "point-bar" deposits because of their characteristic shapes.

Still a third kind of fluvial deposit forms when the stream overflows its banks during flood stage and fills the entire valley or floodplain with water. When the river level again lowers to normal, the trapped floodwaters form shallow bodies of near-stagnant water and the suspended load of fine-grained sediments is permitted to settle out to form relatively extensive mud deposits. Thus, the fluvial sedimentary pattern is one of rather extensive fine sediments disrupted by channel deposits of coarse sediments and crescent-shaped point-bar sand deposits that are more or less randomly distributed throughout the overall accumulation. This description generally fits the lithified sediments found in the basal Chinle members.

In the case of the Shinarump and Moss Back members, the typical sandstone and conglomerate deposits are mainly of the channel and point-bar type of accumulation, and laterally equivalent shales of flood plain origin constitute the interval where coarser-grained cliff-forming deposits are not present. The main site of sedimentation, however, was the point-bar sands, and a considerable amount of plant material including logs, branches, twigs, and leaves that drifted down the Triassic streams was incorporated into the point-bar deposits. This organic material was an important constituent in both the Shinarump and the Moss Back members, for its presence was later to concentrate uranium mineralization of considerable magnitude and importance in these members.

The organic fossil debris creates a reducing environment within the ground waters, causing the precipitation of uranium minerals. The plant material is often replaced by the bright yellow uranium minerals or acts as a nucleus around which the ores are concentrated. Uranium is found to be associated with the fossil plant materials in the point-bar deposits, and exploration for the sometimes valuable minerals is directed toward the location of point-bar sediments in the two members.

Much of the uranium that has been mined from the Colorado Plateau came from the Shinarump and Moss Back members, especially in the Monument Valley and White Canyon districts. The host rock is the plant-bearing sandstone, but occasionally a complete fossil log is found that is entirely altered and replaced by the yellow secondary minerals of uranium, such as carnotite. Such finds are prized by mine operators, as they are valuable bodies of concentrated ore.

The age of the sediment accumulations has been determined from the fossils contained in the formations and is estimated to be about Middle Triassic. The uranium ores have been dated by radioactive age-dating techniques and are found to be not Triassic in age, but much later. Exactly what mechanism provided the uranium-rich fluids to the region is unknown, but it is apparent that the mineral solutions could migrate readily through the porous and permeable sand deposits of the Shinarump and Moss Back members until they came in contact with the localizing organic agents.

The uranium minerals are usually dated as being from the Tertiary, the approximate time during which large intrusive igneous masses were emplaced in several of the ranges of the plateau, such as in the La Sal, Ute, Henry, Abajo, and La Plata mountains. It has been suggested that the intrusions brought the uranium-rich fluids along with the other molten matter and emplaced them in the upper reaches of the earth's crust. The secondary minerals were redistributed through groundwater movements into their present resting sites in the Chinle sands. As this is a plausible explanation for the observed occurrences, it may be considered as a possible mechanism for the ore emplacement.

Others have suggested that the uranium rained down on the Chinle lowlands with airborne ash from volcanic eruptions to the west of the Colorado

Plateau, and that the mineralized ground water trickled down through the Chinle mud to the aquifers of the underlying Shinarump and Moss Back sands.

The main, upper part of the Chinle Formation is made up of varicolored shale, with a few thin beds of fluvial sandstone and dense limestone probably deposited in lakes. The entire Colorado Plateau region must have been a nearly flat plain during Middle to Late Triassic time in order for such extensive deposits of fine-grained sediments to accumulate in river and lake environments. The shale comes in all shades of red, brown, purple, gray, and an occasional pale green, providing pastel landscapes fit for a king wherever the Chinle crops out.

The most lavish colors are found in the broad badlands country of the Cameron-Winslow-Holbrook-Sanders region of north central Arizona, where the formation is extensively exposed at the surface. The region is known generally as the Painted Desert, and a portion of Petrified Forest National Park is set aside to display the multicolored fairyland, along with its contained fossil wood. The thick shale members of the Chinle Formation form high, varicolored slopes above the Shinarump, or Moss Back cliffs in many other parts of the Colorado Plateau, for the formation is distributed around the major uplifted regions in approximately the same patterns as the underlying Moenkopi Formation.

The colorful Vermilion Cliffs north of Grand Canyon are representative of the usual Chinle outcrop outside of the Painted Desert. The red shale slopes surrounding the Defiance and Zuni uplifts are similar in colorful grandeur. Extensive exposures of the Chinle are also prominent in the outer bastions of Canyonlands, where they form a brilliant ring of fire around the inner canyons of Pennsylvanian and Permian strata. The high cliffs guarding the inner canyon world are composed of red sandstone of the next higher unit, the Wingate Sandstone (Early Jurassic age), and the red slopes beneath these upper cliffs are a combination of the lower, brown Moenkopi and the upper, varicolored Chinle formations. Most of the overlooks (with paved-highway access) that permit views of Canyonlands are perched atop the Wingate cliffs, and the outer, red ring of badlands beneath these viewing points are of the Triassic shale formations.

The Chinle Formation is also world-renowned for its content of very abundant logs of highly colored petrified wood. Although the fossil logs are rather widely distributed throughout the Colorado Plateau wherever the formation is exposed, an unusually high concentration of the best material is found near Holbrook, Arizona, and is commemorated and protected by the Petrified Forest National Park. There, hundreds of well-preserved logs, up to about three feet in diameter and at least twenty feet long, are weathering from the very soft Chinle shale beds and lie about on the ground for all to see. But don't touch, or "Smokey" himself will pounce on you. The logs apparently drifted together in large concentrations in the Triassic lakes and floodplains. The cellular pores in the woody tissues were later infilled by silica with its multicolored iron impurities to permineralize (or petrify) the wood.

If the petrified wood is carefully examined, as with a magnifying glass or a microscope, the original woody tissues can be seen as well-preserved organic cellulose materials completely infilled and enclosed in the hard silica. The original organic material has not been replaced by silica (or "changed" to rock), but is simply impregnated with the resistant mineral. The resulting brightly colored petrified wood is a delight to the human observer.

Red beds of Triassic age are present in the San Juan Mountains of southwestern Colorado and are well exposed at several localities. The best places to see these reddish-brown shale and interbedded sandstone layers are just north of the city limits of Durango in the lower Animas Valley, between Dolores and Rico in the west foothills of the range, west of Telluride, and about five miles north of Ouray in the Uncompahgre Valley. The Triassic strata in this region are known as the Dolores Formation and probably include rocks equivalent to both the Moenkopi and Chinle formations, even including a Shinarump-like fluvial sandstone in the lower part.

The Dolores contains much more coarse sandstone than the Chinle, but this is understandable when one considers that this region is much closer to the Uncompahgre uplift source area than the previously described localities. As is usually the case, the coarser material is deposited first and closest to the source because it is considerably more difficult to transport the large and heavier particles than the smaller mud particles. Fossil leaves and bones of primitive vertebrate organisms have been found in the Dolores, which suggests that the climate in this region during Triassic time was rather warm and humid.

CHAPTER SEVEN

JURASSIC DUNES AND DESERTS

All the bleakness of a Sahara Desert returned to the Colorado Plateau during the opening millennia of the Jurassic Period, as the winds blew and the sands drifted across almost every inch of the Colorado Plateau. It is difficult to reconstruct the climate of these times, but it is probable that extreme aridity and high average temperatures were responsible for the great thickness of windblown sand that accumulated during this relatively short interval of geologic time. The change from Chinle stream and lake conditions was abrupt, for the nature of the sediments from shale of the quiet times to cross-stratified sandstones of the windy days is marked by the J-0 unconformity. Sandstone with all the characteristic features of windblown deposits top the Triassic section throughout almost the entire Colorado Plateau Province. The comparison with the Sahara Desert of North Africa does not end with the environmental setting, for the geographic extent of the Lower Jurassic deserts was very comparable to the present-day Sahara. At least the western two thirds of the Colorado Plateau, and adjacent regions, was buried by drifting sand.

Three sandstone formations represent the desert accumulations of this interval of time. The lower of the great windblown deposits is the Wingate Sandstone, which is nearly everywhere a reddish cliff-forming formation that directly overlies the Chinle Formation. The sandstone is usually cross-stratified with the expected steeply dipping individual beds in variably shaped bundles that typify windblown deposits. In a few localities the small-scale, low-angle cross-bedding of stream deposits can be differentiated from the normal texture of the formation. Thus, conditions varied, at least seasonally, within the great southwest desert of Early Jurassic time.

The red Wingate cliffs extend for literally hundreds of miles in a wide, more or less circular pattern enclosing the Canyonlands and the large Monument

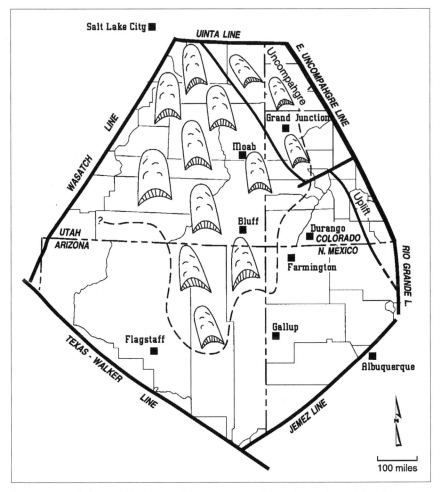

Map showing the known distribution of the Wingate Sandstone of Jurassic age. Note that there is no Wingate Sandstone at the type section at Ft. Wingate east of Gallup, New Mexico, and the massive sandstone crosses over the Uncompahgre uplift near Grand Junction, Colorado.

Upwarp, effectively protecting the inner canyon country from invasion. The cliffs rise several hundred feet throughout the region, with very few breaks in the pattern. Consequently, access to the lower country must be literally carved from the vertical cliffs of the Wingate, and then it is usually crossed only on foot or by four-wheel-drive vehicles. For this reason one must sometimes travel many tens of miles, or even farther, to locate an access road through the Wingate cliffs. Then the traverse is often so precarious as to discourage even the most experienced desert traveler. Although the Wingate Sandstone averages only three hundred feet in thickness in this region, it is an effective

Aerial view of Lake Powell where it fills the former site of Glen Canyon, backed up behind the Glen Canyon Dam in northern Arizona and southern Utah. The light-colored rocks are the Glen Canyon Group, including the Navajo-Kayenta-Wingate formations, in descending order, at their type section. The large round mountain in the distance is Navajo Mountain, which is a structural dome, a laccolithic igneous intrusion, capped by strata of the Cretaceous Dakota Sandstone.

barrier to travel onto almost any of the major uplifts.

The upper of the two prominent windblown sandstone formations is the Navajo Sandstone of Middle Jurassic age. It is separated from the Wingate by extensive fluvial sandstones of the Kayenta Formation. The combination makes a tripartite group of closely related formations called the Glen Canyon Group, named for spectacular exposures of the threesome in Glen Canyon of the Colorado River. Glen Canyon is now largely inundated by Lake Powell. Of the three formations, the Kayenta is the more friendly to travelers, because it usually erodes into relatively gentle ledgy slopes.

The Kayenta Formation dissects the Glen Canyon Group by forming a ledgy slope between two massive cliff-forming sandstones. It differs also in that it is composed of distinctive small-scale cross-stratification with low individual dips of a fluvial origin. It is likely that the stream deposits were not formed simultaneously across the entire Colorado Plateau, but the formation invariably occurs about midway in the otherwise massive group. Because conditions were somewhat better in Kayenta time for the preservation of tracks, it is sometimes possible to find dinosaur tracks on the bedding planes. A handy place to see these is just west of Moab, Utah, along the highway leading along the Colorado River to Potash. This is also an excellent place to view the entire Glen Canyon Group in typical exposures.

The Navajo Sandstone is more like the Wingate in that it forms steep, rounded cliffs that are impassable or nearly so at many exposures, especially where the strata are upturned surrounding an uplift such as the Monument or the San Rafael flexures.

The Glen Canyon Group, then, consistently forms impassable cliffs or hogbacks wherever it crops out, producing "reefs," which effectively block access to the uplifts. A prime example of such a barrier hogback is displayed at Capitol Reef National Park, where the Glen Canyon Group flanks the Waterpocket Fold, a sharp flexure in the earth's crust extending for about 150 miles in a north-south direction a few miles west of Hanksville, Utah. Other spectacular examples are the San Rafael "reef" to the northeast, and Comb Ridge west of Bluff, Utah

The Wingate and Kayenta Formations grade from a sandstone to a thicker siltstone unit toward the southwest of the type section at Kayenta, Arizona, and grades into another sandstone known as the Moenave Formation in the vicinity of Tuba City. The Moenave and the Kayenta formations have yielded the bones of dinosaurs and of a mammal-like tritylodon that date the formations as probably Jurassic in age. Geologists with the USGS have discovered plant spores in the Glen Canyon Group that verify its Jurassic age, after decades of controversy. Consequently, the Wingate, Moenave, Kayenta, and Navajo formations are now considered to be of Jurassic age.

The great Navajo desert extended almost across the entire width and breadth of the province, cloaking the surface with dune sands interrupted only locally by playa lakes (desert lakes that dry up during the dry seasons). The resulting sandstone is a classic example of windblown deposits, having large-scale, high-angle cross-stratification in conspicuous wedge-shaped sets wherever the formation is exposed. The Navajo Sandstone is usually white

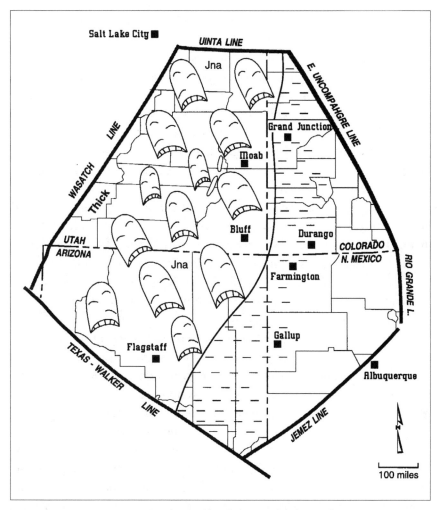

*Map showing that the Navajo Sandstone of Jurassic age is present
only in the western half of the Colorado Plateau.*

and composed of fine, uniform grains of almost pure quartz sand. In some
localities, such as in the Four Corners area, it also contains thin beds of fresh-
water limestone that extend for only perhaps a city block or two—moot evidence of the presence of ephemeral lakes in the Jurassic deserts.

By far the best place to see the Navajo Sandstone is in and around Zion
National Park in southwestern Utah. It is the massive cliffs of the Navajo that
form the precipitous canyon walls in Zion as the formation attains its maximum thickness of at least two thousand feet. Where the walls are vertical and
massive, it is difficult to see the wild array of cross-stratification features, but

where the sandstone is weathered into rounded slopes it becomes apparent that the wind demons were busy when the Navajo desert was in production. The cross-stratification spectacle is well displayed near the eastern park boundary along the highway, where steeply inclined, highly irregular, and contorted sandstone beds are the rule rather than the exception. It is obvious that the Navajo Sandstone is a giant frozen pudding of "fossil" sand dunes, when viewed in such a locality. Although the Navajo is normally less resistant to erosion than the underlying Wingate Sandstone, it forms respectable and impressive cliffs, canyons, buttes, and mesas in the Zion country unparalleled by any other formation.

The Navajo Sandstone is also prominently displayed across much of the Navajo Indian Reservation of northeastern Arizona, from which it gets its name. It surrounds the great flexures of the Colorado Plateau such as the Monument Upwarp, the San Rafael Swell, the Waterpocket Fold, and others. The formation pinches out eastward before reaching the San Juan Mountains and Uncompahgre Plateau, but otherwise is a producer of exceptional scenery throughout the plateau country.

Immediately overlying the Navajo Sandstone and the J2 unconformity is a maverick body of light-colored sandstone that is nearly indistinguishable from the Navajo Sandstone. The newly named formation is the Page Sandstone, named for exposures near the town of Page in north-central Arizona. The key to recognition of the Page Sandstone is that it overlies the erosional surface at the top of the Navajo. One must be cognizant of the presence of the separate formation, or it is commonly included in the upper Navajo Sandstone in the field. Justification for naming the new sandstone formation is that it overlies the J2 unconformity, and interfingers with the Carmel Formation. The Page Sandstone extends from Page eastward to Navajo Mountain, north to Escalante, and west to near Kanab, Utah. Its extent southward into Arizona is unknown due to the nature of surface exposures. Because the Page is the same age as the lower Carmel Formation, it is included in the San Rafael Group, despite its physical similarities to the Navajo Sandstone.

Much as in the Permian Period before it, the red beds of the Triassic and the sandstone of the Jurassic are responsible for providing the Colorado Plateau with much of its unique color and beauty. In fact, the Permian through Jurassic formations are often very difficult to distinguish if their relative stratigraphic position is not understood. For example, the Permian buttes and spires composed of the Organ Rock Shale overlain by the De Chelly Sandstone in Monument Valley are duplicated like carbon copies by the buttes of the younger

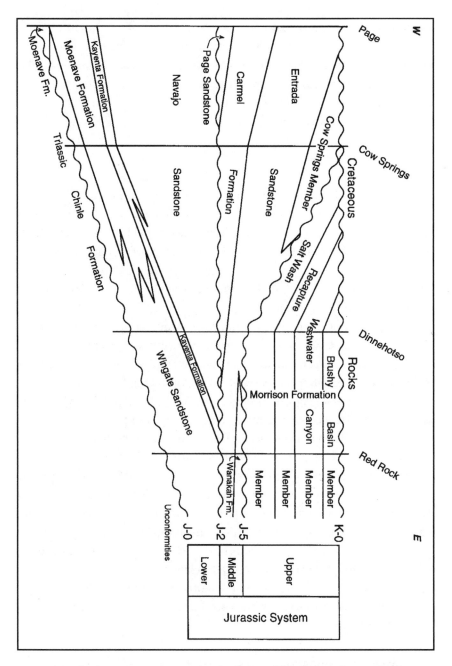

Generalized cross section showing the relationships of the various rock units of Jurassic age across the northern Navajo Indian Reservation, from Lake Powell east to the Four Corners. The wavy lines indicate regional unconformities (J-0 to J-5) that serve to divide the various units on a regional basis.

Chinle Shale-Wingate Sandstone sequence in several places, notably around the periphery of Canyonlands. The combination of the red Permian and Triassic strata provides hundreds of feet of spectacularly colorful red rocks that defy description, splashing flaming hues across thousands of square miles of desert. But the Colorado Plateau is not yet completed, for the Late Jurassic formations have not yet been accounted for.

Late Jurassic Red Beds

A very sharp break in the bedding occurs at the top of the Navajo Sandstone, where the cross-stratification of the windblown sands is sharply truncated by erosion. The red beds of the overlying Carmel Formation are always in sharp contact with the Navajo, with no evidence of gradation having occurred. This erosional surface, the J-2 unconformity, is a good place to begin again in our geologic story, as the formations above the break are definitely known to be Middle Jurassic in age.

A large part of the known Jurassic formations on the Colorado Plateau belong to the San Rafael Group, which is named for extensive exposures of this part of the geologic column on and near the San Rafael Swell of central Utah. The group is composed of alternating sequences of red and brown mudstone and shale, with intervening light-colored beds of sandstone. The origin of the group is marine, or some closely related environment, along the western parts of the province, especially in the Zion Canyon and San Rafael Swell regions. It grades generally to continental sediments in the central and eastern Colorado Plateau.

To the far west, in the vicinity of the Great Basin and Salt Lake City country, the entire section is represented by marine fossiliferous limestone of the Twin Creek Formation. It is apparent, then, that the San Rafael Group represents a lateral gradation of environments from totally marine on the west to totally continental on the east, with intermediate marginal marine settings distributed through the central Colorado Plateau.

The Jurassic Period closed after the return of continental deposits called the Morrison Formation, which are distributed throughout almost the entire Colorado Plateau.

The Carmel Formation

The oldest formation in the San Rafael Group is the Carmel, named for Mount Carmel, just east of Zion National Park. In that region, the Carmel

Formation is composed mainly of limestone and shale, with interbeds of gypsum, that contain a marine mollusk fauna that dates the formation as Middle Jurassic in age. The same fossils occur in the lower part of the formation in the type region of the San Rafael Group. Throughout the remainder of the Colorado Plateau, the Carmel is a series of red or reddish-brown mudstone and siltstone that contain no age-definitive fossils.

The red facies of the Carmel is most likely of a mud-flat origin in the central part of the region around Canyonlands and the Monument Upwarp, but it grades eastward to a more continental lowland type of deposit with some evidence of stream sedimentation having mixed with the mud flats and floodplains. The source of the red sediments was probably to the east of the main Colorado Plateau, perhaps in the Uncompahgre uplift region, but the formation thins in that direction from a maximum of about 650 feet on the southern San Rafael Swell to the pinch-out of the formation along the Uncompahgre front near Gateway, Colorado.

A peculiar feature of the Carmel Formation in the general vicinity of Moab and Bluff, Utah, is large-scale contorted bedding. The irregular contortions in the well-bedded facies often include the entire formation as a single unit, suggesting that whatever caused the unusual crinkling of the strata occurred after the entire formation was deposited. The irregularities usually also involve the basal beds of the overlying formation, the Entrada Sandstone, further implying that the higher formation was also present at the time the beds were disrupted. Since the upper parts of the Entrada are not disturbed, it is likely that the phenomenon occurred during its deposition. All this seems to indicate that the contortions were formed after deposition of the Carmel and after some loading by the overlying beds occurred. It must have happened before the mud and sand had been firmly compacted and cemented into well-indurated rocks.

About the only thing that could cause such widespread disturbance of the bedding would be gravitational slumping down some sort of slope, but what could cause such a slope to develop? Probably a slight rejuvenation of several preexisting structures tilted various local areas, creating the appearance of a more regional condition. At any rate the wild contortions are a typical characteristic of the Carmel over large areas, and can be seen in such widely separated places as Arches National Park and the Bluff to Red Mesa region along the Utah-Arizona border. The Carmel is locally considered to be the Dewey Bridge Member of the Entrada Sandstone in Arches National Park, for convenience in mapping.

The Entrada Sandstone

The second formation from the base of the San Rafael Group, the Entrada Sandstone, forms distinctive and unique topographic features wherever it crops out. In the San Rafael Swell region, where it was named, it is composed of a pale reddish-brown earthy sandstone and siltstone deposited in massive or uniformly bedded layers. The formation was undoubtedly formed in an aqueous environment.

The relatively soft-weathering formation is not very different from the underlying Carmel Formation in this region, except that at many localities it forms strangely shaped irregular knobs known as "hoodoos." One of the best developments of this strange weathering phenomenon is at Goblin Valley State Park just east of the San Rafael "reef" a few miles north of Hanksville, Utah. There the earthy facies of the Entrada is weathered into grotesque figures of unending variation that stand like a comical army in formation awaiting some unknown event. The peculiar little goblins are fascinating to those who enjoy artistic but useless forms. Because of its unusual weathering characteristics, the earthy facies of the formation is often referred to as the "hoodoo Entrada." This facies ranges upward to about 850 feet in thickness.

To the east and south of the San Rafael Swell, the Entrada grades rapidly into a massive, cross-stratified sandstone of prominent proportions. Although the earthy facies was undoubtedly formed in a marine environment, the sandy facies is generally thought to have been deposited as dune sands much like the Navajo below. The large-scale cross-stratification is not always in evidence, however, and the formation is a massive and uniformly bedded unit in many localities. Regardless of its origin, the sandy facies of the Entrada invariably forms massive rounded-to-overhanging cliffs and spectacular scenery, in the Moab region in particular.

To the north of Moab, the Entrada, along with the underlying soft, contorted beds of the Carmel, erode to tall, irregular-shaped spires and statuettes along the entry road into Arches National Park. Then, along the flanks of Salt Valley in the heart of the park, the Entrada weathers into menacing walls and buttresses that resemble giant multiple fins where the massive sandstone has been highly fractured by the intrusion of the Paradox salt. Because of its massive nature, the sandstone cliffs spall off in rounded, overhanging, natural alcoves. When undercutting attacks one of the thin sandstone ribs, usually at a water seep, it often breaks through the thin wall of sandstone to form a hole—a natural arch results. The arches are so well developed in the

Salt Valley region that Arches National Park was created to protect and display the natural erosional peculiarities. Others are scattered throughout the greater Moab area, but none are so outstanding as in the Park. The Entrada Sandstone also forms prominent reddish cliffs of exceptional beauty in the region crossed by U.S. Highway 160 between Moab and Monticello, Utah.

In the Four Corners country and on into the Navajo country of northeastern Arizona, the Entrada is not so bold in its exposures, but it still forms cliffy topography between soft-weathering, slope-forming formations above and below. In this region it is more reddish than in most other localities, and locally includes interbeds of red siltstone. The Cow Springs Sandstone of the Black Mesa basin is now considered to be a bleached, light colored facies of the Entrada Sandstone.

East of the Four Corners and southeast of Moab, the Entrada retains its massive, highly cross-stratified nature, but it changes in color to a clean white. This uncolored occurrence is present in the exposures around Dolores, Colorado, and extends through intermittent exposures into the San Juan Mountains of southwestern Colorado. There, the Entrada Sandstone plays the lower part of a duet of white massive sandstone cliffs, as another very similar formation overlies the Entrada but is separated from it by a thin red unit. The white sandstone doublet forms a distinctive marker in the Animas Valley immediately north of Durango, Colorado. The Entrada Sandstone in the clean, white facies is most probably of windblown origin.

The Curtis Sandstone

A reorientation of the marine byways occurred late in Jurassic time following the close of the Entrada scene. Until this time, the seaway of the Mesozoic lay mainly to the west of the Colorado Plateau and extended generally in a north-south direction through western Utah and northward into Wyoming and Montana. A short interval of time separated the Entrada from the next-overlying formation, the Curtis Sandstone, for an erosional surface intervenes that truncates minor folds in the Entrada bedding in the San Rafael Swell region. This is the J-3 unconformity. When sedimentation again commenced, the regime had changed almost completely and the Curtis sea stretched away to the east across northern Utah and southern Wyoming. The only sedimentation on the Colorado Plateau of any importance occurred on the San Rafael Swell, where a white marine sandstone overlies the Entrada earthy beds with a thin conglomerate marking the base of the new formation. The light-colored sandstone

---

of the Curtis is usually either massive and structureless or regimented into horizontal thin beds. Beds of light-gray shale may be present. The sandstone contains significant amounts of glauconite, a green iron potassium silicate mineral that forms only in relatively shallow marine environments. Marine mollusks and microfossils date the Curtis as definitely late Middle Jurassic.

The Curtis Sandstone reaches a maximum thickness of some 250 feet in the northern San Rafael Swell, but it thins rapidly to the south and east. It is unknown east of the Green and Colorado rivers, except in northernmost Utah and Colorado in the Uinta Mountains, as it interfingers into the basal Summerville red beds just east of the Swell. It thins rapidly toward the south and pinches out in the vicinity of the Henry Mountains south of Hanksville. Consequently, the Curtis Sandstone is not recognized even in Canyonlands country.

A thin interval of dark limestone and white gypsum occurs in the same stratigraphic position between the Entrada and the Summerville Formations in southwestern Colorado and northwestern New Mexico. This unusual unit, called the Todilto Formation, was deposited at approximately the same time as the Curtis. There is no evidence that the two formations were ever connected. In fact, most geologists who have studied the Todilto feel that it is not even marine, but a fresh-water lake deposit. The evidence to support that interpretation is meager indeed, but there is no reason to believe that it is marine, either. No diagnostic fossils have ever been found in the formation, a negative line of evidence that weakly supports the theory of a nonmarine environment of deposition. Several specimens of rather large fossil fish have been found in the formation, but they are similar to the salmon in that they are believed to have lived parts of their life cycles in both marine and fresh-water worlds. Gypsum occurs locally in the San Juan Mountains near Ouray, Colorado, and is well developed in thick beds in the San Juan basin in northwestern New Mexico. The evaporitic sediment can precipitate from either fresh or salt water and is therefore not very informative.

If the Todilto is lacustrine (lake) in origin, then the lake was a rather extensive body of fresh water that was completely disconnected from the Curtis sea to the north. If it was formed in a marine embayment of some kind, there is no evidence to suggest where the main body of seawater lay. The presence of significant amounts of gypsum certainly implies that the climate was arid, and since the evaporites generally occur in the upper parts of the formation, it is altogether possible that they represent the drying-up stages of an extensive body of water, be it fresh or marine. The gypsum beds of northwestern New Mexico are intimately interlaminated with thin dolomite

layers in a cyclic succession. It has been suggested that the deposits originated during climatic cycles that were controlled by Jurassic sunspot cycles.
(Anyone care to argue the point?)

The Last Red Beds

The Curtis sea retreated northward in late Middle Jurassic time, followed
by extensive intertidal mud flats of the Summerville Formation. The chocolate-
colored siltstone and mudstone is almost universally horizontally thin-bedded, with some thin, white sand beds punctuating the sequence. The formation
is very distinctive in appearance because it is browner than the other red beds of
the San Rafael Group and forms ledgy cliffs of the earthy clastic rocks.

The Summerville is an interesting formation because it is intimately related to light-colored sandstone formations. Although it overlies the Curtis
Sandstone in the San Rafael Swell, the Summerville grades into the marine
sandstone to the north and interfingers with basal Curtis beds in the eastern
San Rafael region. From these relationships it is apparent that the Summerville
represents the tidal flats marginal to the Curtis sea as it retreated northward
into Wyoming. That the Summerville was deposited in tidal-flat conditions
is indicated by the abundant presence of parallel and symmetrical ripple
marks and mud cracks that require shallow waters and occasional desicca-
tion to form. Also, the thin, horizontal bedding is very difficult to produce
in other environments. The Summerville mud flats were involved with windblown sandstone in the Kaiparowits area, where the Romano Sandstone is
an apparent equivalent of the Summervile.

The Summerville Formation is widespread on the central Colorado Plateau, in spite of its unorthodox relationships. It is thickest on the San Rafael
Swell, where it is over 325 feet thick. It thins by grading into the Curtis toward
the north until the Curtis is considered to be a thin member of the Stump
Formation in the Uinta Mountains of northeastern Utah. It is missing in the
Moab region, although it may be partially equivalent to the Moab Tongue of
the Entrada Sandstone. Over all its geographic extent, the Summerville averages
about seventy-five feet in thickness and rarely exceeds one hundred feet.

Formations Previously Thought to be Summerville Equivalents

A series of red beds, formerly identified with the Summerville Formation, interfingers with the Bluff Sandstone in the Blanding and Bluff, Utah,

country. This unit, which occurs above the J-5 unconformity, has recently been named the Tidwell Member of the Morrison Formation. More recently, geologists have included the Bluff Sandstone with the overlying Morrison Formation. Another pair of prominent sandstone formations known as the Winsor Sandstone of south central Utah and the Cow Springs Sandstone of north central Arizona appear to be windblown equivalents of all or at least parts of the Entrada Sandstone.

In the San Juan Mountains the Entrada and Todilto are overlain by the Wanakah Formation, forming a prominent slope between the massive white cliffs of the Entrada and Junction Creek Sandstones. The Wanakah, named for a mine above Ouray, Colorado, was originally called the Summerville, however, the two formations are now known to be of separate origins. The Wanakah is now mapped across the Uncompahgre Plateau and into the Four Corners region at the stratigraphic horizon previously called Summerville. Windblown sandstone beds of the Bluff and Junction Creek Sandstone, that occur above the Wanakah Formation in the San Juan Mountains and in the Four Corners country were previously included in the San Rafael Group even though they are not present in the type area. It is not really important how the sand bodies are classified so long as their presence and significance are understood. Regardless of which paragraph they occupy in the book of historical geology, the Bluff Sandstone of the Four Corners area and the Junction Creek Sandstone of the San Juan Mountains form prominent massive cliffs in their respective areas. They also form the uppermost of the significant windblown sandstone formations and top the last of the great red beds that typify the Colorado Plateau.

Of Lands and Dinosaurs

Strictly continental conditions settled over the entire Colorado Plateau and southern Rocky Mountains in latest Jurassic time, bringing with it the deposits of the Morrison Formation. These accumulated sediments consist mainly of pale green, gray, and reddish mudstone and sandstone of definitely lake and stream origin. In some places thin limestone beds occur in the Morrison that contain the fossil reproductive bodies of green algae (charophytes) that live strictly in fresh water. The Morrison Formation usually includes a lower member of stream-deposited sandstones and an upper member of floodplain and lake mudstones. In some localities the formation also contains other sandstones within the main body of the unit, providing further means of subdividing the formation into members. Regardless of the vertical position within the formation, the Morrison consists of fluvial sand deposits interspersed with lake and floodplain mud and silt; the relative proportions vary within the formation to provide the vertically segregated concentrations of the different lithologies. There are five distinctive members recognized in the Morrison of the Colorado Plateau, and they are all present and well developed in the immediate vicinity of the Four Corners area.

The five members, from the base upward, are: The Tidwell Member, fine-grained stream deposits that directly overlie the J-5 unconformity; the Salt Wash Sandstone, a strongly fluvial deposit; this is overlain by the Recapture Shale Member; another fluvial sandstone known as the Westwater Canyon Sandstone Member; and an upper mudstone unit called the Brushy Basin Shale Member. Of the five, the Brushy Basin Shale Member is the most widespread in occurrence, being present over most of the Colorado Plateau country and perhaps extending even to the type section of the Morrison Formation just west of Denver, Colorado.

The Tidwell Member is a light-colored mudstone that extends over most of the Colorado Plateau directly overlying the J-5 unconformity. Recognition of the newly named member and the J-5 unconformity readily explains the relationships between the Summerville below and Morrison Formation above.

The Salt Wash Sandstone Member directly overlies the Tidwell Member above the base of the Morrison, and is also nearly ubiquitous on the plateau. It generally thickens toward the west and southwest of the Four Corners.

The Recapture Shale and the Westwater Canyon Sandstone, are relatively restricted in distribution, being found mainly in northeastern Arizona and northwestern New Mexico. These relationships are useful in understanding the generalized stratigraphy of the Colorado Plateau. When studied in detail, however, the members are highly variable laterally, and grade into one another with the greatest of ease.

The thickness trends of the various members and the distribution of the coarser grain sizes in the sandstone beds suggest that there were two different areas supplying sand and mud to the Colorado Plateau in Morrison time. The Salt Wash Member thickens and becomes coarser in grain size toward the southwest of the Four Corners, and the fluvial cross-stratification dips toward the northeast on the average. All this points to a source area toward the southwest, perhaps somewhere in central Arizona. Detailed comparisons of the mineralogy of the sediment particles suggests that the Salt Wash and the Brushy Basin members had a common source.

In contrast, the Recapture and Westwater Canyon members thicken and become coarser southward from the Four Corners, and their similar mineral content suggests that their common source area must lie to the south. Considerable amounts of volcanic ash deposits are present in all the members, but are most heavily concentrated in the Brushy Basin Member. Volcanic ash can remain suspended in the mobile atmosphere sometimes for weeks or even months and the fine-grained material is so readily transported by even the most sluggish streams. It is not at all certain as to the location of the parent volcanoes or even their general direction from the Four Corners.

The many similarities in composition and origin of the Shinarump and Moss Back sandstones to the widespread Salt Wash Member make it not at all surprising that the basal Morrison sandstone beds contain significant reserves of uranium. As in the case of the Triassic deposits, the Salt Wash contains most of its ore in the immediate vicinity of concentrations of organic debris in the point-bar deposits, and the trashy beds containing considerable amounts of plant material have been prospected with enthusiasm. The geographic distribution of the Salt Wash ores is unique, however, since they are localized in the "Uravan Mineral Belt." This belt of mineralization lies east of Canyonlands in a more or less circular pattern surrounding the La Sal Mountains. The deposits are especially rich to the east and south of the range in the vicinity of the Paradox, Gypsum, and Lisbon valleys. Despite the limited distribution of the mineralization, the Salt Wash Member produced significant amounts of ore and has been a world-wide leader in production and reserves.

The area north and east of Grants, New Mexico, has more recently emerged as the largest uranium-mining district in the United States. Here, the ore-producing members of the Morrison Formation are the Westwater Canyon Member and a local member known as the Jackpile Sandstone that occurs above the Brushy Basin Member. The combined reserves of these deposits is estimated at 500 million pounds of U_3O_8 "yellowcake." As elsewhere on the Colorado Plateau, the ore occurs in fluvial sandstone bodies in close association with fossil plant debris, but unlike the other deposits, the uranium occurs mainly as the gray to black-colored oxide known as uraninite. The mineralization in the upper Morrison Jackpile Sandstone Member is closely associated with the overlying Dakota Sandstone of Cretaceous age, and it is probable that the uranium-bearing groundwaters were derived from the highly carbonaceous Dakota. At any rate, the mineralization took place several million years after deposition of the host rocks.

Dinosaurs had long since taken over the earth by Morrison time, but their remains are remarkably scarce in the underlying formations. Perhaps this is because much of the Triassic and Jurassic section was deposited in marginal marine lagoons and tidal flats that were relatively inhospitable, rather than totally stream and lake environments such as are represented by the Morrison Formation. Whatever the cause, the Morrison is the only formation on the Colorado Plateau to contain significant numbers of fossil dinosaurs. And it is a veritable bone yard. Dinosaur bones, usually preserved by the infilling of the porous microstructure of the bone by silica, much like the petrified wood of the Chinle, can usually be found almost anywhere that the Morrison is exposed. The bones are usually found separately in partially eroded projections from the shale or sandstone outcrops. Instead of looking like an enormous dog bone they usually appear more like a brownish boulder weathering out of the softer, sedimentary layers. In a number of localities, however, the dinosaurs are present in very nearly complete condition and in large concentrations of individual animal skeletons.

The best-known of these concentrations of fossil dinosaurs is in Dinosaur National Monument, near Vernal, in northeastern Utah. A great many specimens have been taken from these quarries and many more still exist in the rock. They are found in a hard sandstone bed of obviously fluvial origin, where they were undoubtedly carried by flowing streams after their deaths. Even at that, the area must have been a dinosaur "center," for such large numbers of specimens are unknown in most other localities. Many early collections were made from this quarry, which supplied dozens of dinosaurs to the large eastern museums. The very tedious and delicate job of extricating the bones from their rock matrix is still in progress at the Monument, where the paleontologists are baring the fossils to view but leaving them in their natural positions in order to display the nature of fossil occurrences.

Many other fossil-collecting localities are known. For example, numerous specimens have been quarried from the up-turned Morrison outcrops flanking the northwesterly plunging nose of the Uncompahgre uplift. The quarries occur just north of Colorado National Monument near Grand Junction, Colorado, and in Rabbit Valley near the Utah-Colorado border.

Another, more recently discovered dinosaur quarry is in the Morrison Formation along the west flank of the San Rafael Swell south of Price, Utah. At the latter occurrence, the late Wm. Lee Stokes of the University of Utah has quarried the fossil beasts and sold them to other universities and museums on a nonprofit basis.

The Morrison dinosaur fauna includes many of the well-known varieties cherished by museums and their visitors. The carnivorous types such as Allosaurus are commonly found in the Upper Jurassic deposits along with such herbivorous monstrosities as Brontosaurus and Diplodocus, whose only danger to other, contemporaneous animals was that they might accidentally step on them. These herbivorous dinosaurs grew to about eighty feet in length and weighed many tons, yet their brain cases were significantly small, suggesting that they were blimp-sized imbeciles. Several other dinosaurs, of lesser repute, such as Stegosaurus and Brachiosaurus, have been found in this prolific formation also. Truly, the pea-brained monsters dominated the life and landscape of the Morrison lowlands as the curtains fell across the Jurassic scene.

CHAPTER EIGHT

THE LAST OF THE GREAT INLAND SEAS

THE CRETACEOUS PERIOD

The Cretaceous Period dawned over the Colorado Plateau with very little pomp and ceremony. In fact, it is often very difficult to decide where the lowland deposits of the Morrison Formation leave off and the lowland deposits of the Lower Cretaceous Burro Canyon (or Cedar Mountain) Formation begin. In a general sense, however, the basal Cretaceous deposits can be distinguished from the Morrison varicolored upper shales because they contain an increased amount of sandstone and conglomerate beds, which have the habit of being quite resistant to erosion, thereby forming prominent cliffs. The basal Cretaceous sandstone is generally light brown, but it is commonly interspersed with light-green and pale-purple shale of a decidedly Morrison-like appearance. The distinction is sometimes difficult when detailed relationships must be considered. The Lower Cretaceous sediments were also deposited in much the same manner as the Upper Jurassic Morrison Formation, further complicating the situation. Both units were formed mainly in continental environments consisting of low-gradient streams meandering through relatively flat lowlands and occasional lakes.

The Lower Cretaceous landscapes varied only slightly from the previous episode. Indeed, the distinction in age was possible only after considerable study of the microfossils. Tiny bivalve arthropods known as ostracods, resembling Lower Cretaceous fresh-water faunas from other parts of the world, were present in parts of the Burro Canyon Formation. Cliffs of the Burro Canyon extend throughout much of the partially dissected uplands of east central Utah and southwestern Colorado, where the formation ranges upward to over 250 feet in thickness. Because physical correlations are not possible due to the nature of the exposures, very similar rocks west of the Green River were renamed the Cedar Mountain Formation.

The long reign of the continental environments closed as a relatively brief interval of erosion scarred the Burro Canyon and Cedar Mountain deposits. The resulting unconformity is usually obvious and marks the approximate changeover from Early to Late Cretaceous time. The erosional surface sometimes carved away the entire Burro Canyon Formation, permitting Upper Cretaceous sediments to come to rest on Jurassic or even older strata in some areas, such as toward the southwest in central Arizona. The first Upper Cretaceous deposits, the Dakota Sandstone, are also composed of sandstone and carbonaceous mudstone that may form prominent cliffs. Consequently, the Lower Cretaceous beds often merge with the basal Upper Cretaceous deposits into a single cliff when viewed from a distance. However, the regime changed from continental to marginal marine environments as the mid-Cretaceous passed into history, and the later sandstone beds are usually distinctive and significant.

The Seas Return

With the advent of Late Cretaceous time, the sea once more marched across the width and breadth of the Colorado Plateau, but for the last time. As might be expected, the initial deposits represent the beaches that migrated across the region just ahead of the sea. Beach sands were not the only indication of the passing shoreline environments. Sand-bar, swamp, and lagoonal deposits are also to be found in the basal formation known as the Dakota Sandstone. As a result of these several interrelated environments and their deposits, the Dakota is often quite variable in detailed lithology from one locality to the next. The Dakota Formation is, however, one of the most widespread units in the West, or at least it appears to be. It is recognized throughout vast regions of the Midcontinent, where it underlies the plains and furnishes artesian waters to the otherwise rather dry region. In fact, the name was derived from the Dakota region of the northern plains and extended by early geologists across the Rocky Mountains because of similarity of rock types at this horizon wherever it is exposed. It is quite possible that the Dakota Sandstone is composed of different disconnected sand bodies in its various occurrences, and it is highly probable that the formation is of somewhat different ages in different localities. However, it is still considered to be a single formation throughout the middle stretches of the United States. It represents deposits along the advancing Cretaceous shoreline in spite of its local age or lithologic idiosyncrasies. Its widespread distribution sympathizes with the overlying marine formations of Upper

Cretaceous age, thereby testifying to the vast extent of the Upper Cretaceous seas and the distances the Dakota beaches traveled.

In the Colorado Plateau province the Dakota Sandstone is typically divisible into three distinctive parts, which may or may not be present at any particular locality. A lower, cliff-forming unit usually consists of a conglomerate or highly cross-stratified coarse sandstone of light-brown to gray color. The member probably represents the initial deposits on the eroded Lower Cretaceous land surface, incorporating the debris left behind as the mid-Cretaceous unconformity was overlapped. As the initial onslaught of the shoreline processes subsided, a middle unit consisting of carbonaceous shale and coal formed, as marsh and lagoonal environments struggled with the new elements for domination of the region. This middle, usually soft-weathering, member gives way upward gradually to another sandstone, this time a typical beach deposit with fine, well-sorted grains in large-scale, low-angle beachlike cross-stratification. The combined effect of the three subdivisions totals only some 100–150 feet of sediments, but the significance of the Dakota Sandstone as a transgressive shoreline deposit more than offsets its physical ineptitude.

The age of the Dakota Sandstone is not well established. As is often the case, beach deposits and their associated marsh and lagoonal sediments are almost devoid of good datable fossils. Only in the uppermost marine unit have clams and cephalopods been discovered, and then only rarely. Consequently, the formation is known to be basal Upper Cretaceous in age where these distinctive faunas are found, but elsewhere it is uncertain whether or not the age is similar, or to what extent it may vary.

The sandstone beds of the Dakota are often quite porous and permeable, and consequently the formation makes a good reservoir for oil and gas where it is fairly thick and devoid of argillaceous (muddy) sediments. Considerable amounts of petroleum, both oil and gas, have been produced from Dakota reservoirs in the San Juan basin of northwestern New Mexico. Production is probably mostly from offshore sand bars and related sand accumulations where much of the muddy sediments that plug up the porosity were winnowed out at the time of deposition. High-quality oil has been produced from the Dakota Sandstone in the Four Corners area since its discovery in 1923 at Hogback Field, but only in recent years have oil and gas fields been fully developed in the deep central portion of the San Juan basin. As might be expected, low petroleum prices and the high cost of deep drilling have made the deeper production uneconomical until the late 1970s. The deep Basin Dakota field is now one of the largest in the Southwest.

Aerial view of Durango, Colorado. Badlands topography in the upper Cretaceous Mancos Shale is capped by Mesa Verde sandstone in the middle distance just beyond town. The peaks on the skyline are in the La Plata range and are composed of sedimentary strata intruded by igneous sills of Tertiary age.

The Black Mud of Mancos

As the seas deepened and the Dakota shoreline marched into the distant horizon, black mud began oozing and trickling down from the Upper Cretaceous waters. The accumulated sediments are the extreme opposite of red beds in significance, for they require entirely different circumstances for their formation. The dark gray and black colors result from either relatively high proportions of organic debris in the mud or the presence of iron pigment in the ferrous, or deoxygenated, state. In either case, the site of deposition must be very low in oxygen so that the sediments are reduced rather than oxidized, or the dark pigments will be destroyed or turned to red.

Environmental conditions that form black sediments are generally deep, or at least stagnant, waters in which the oxygen content of the water is not readily replenished. Such an environment causes the iron compounds to be reduced to a state in which the oxygen-to-iron ratio is low. It also prohibits or retards the activities of scavenging organisms, which would remove and utilize much of the dark organic materials that also contribute to the dark gray and black hues. Of course, if the organic substances were exposed to oxidizing conditions they would be destroyed and lost as agents of coloration. So seawater of relatively stagnant character inherited the Colorado Plateau, as well as the interior United States, during Late Cretaceous time.

The black shale that overlies the Dakota Sandstone is prominent over large areas of the province. Perhaps they are best developed in the vicinity of Mesa Verde National Park in southwestern Colorado; at least that is where the type section is located. There, in the vicinity of Mancos, Colorado, the unit attains a thickness of about two thousand feet and forms massive gray slopes that weather into badlands-type topography surrounding the main park area.

The Mancos Shale, as it is called, surrounds the entire San Juan basin of northwestern New Mexico and southwestern Colorado. In and surrounding the townsites of Cortez, Mancos, and Durango, the Mancos Shale forms the high, steel-gray slopes and cliffs that dominate the local scenery. The eerie gray cliffs swing on to the southeast and border the San Juan basin to the vicinity of Cuba and Albuquerque, then go westward to Grants and Gallup, New Mexico, before swinging northward again toward Shiprock and finally closing the loop at Cortez.

Very similar dark gray cliffs of Mancos Shale surround the Black Mesa basin, which lies between the Defiance and the Kaibab uplifts of northeastern Arizona. In both cases the Mancos dips centripetally toward the basin center and produces a "bathtub ring" around the large, structurally depressed basin. Huge gray shale slopes of the Book Cliffs, that form the southern margin of the Uinta basin of northeastern Utah, appear identical to the Mancos Shale of the Mesa Verde region. There the Mancos forms gray slopes and cliffs that extend all the way from Price, Utah, to Grand Junction, Colorado.

From this discussion it can be readily seen that the younger formations, such as the Mancos Shale, are found mainly in the structurally depressed basins, where they are relatively safe from the effects of erosion. This also gives a lifeless gray complexion to the basins in contrast to the fiery red hues of older rocks on the uplifts. In spite of the dull and innocuous appearance of the Mancos Shale, it must be avoided like the plague when it is wet. Dirt roads are changed from hard surfaces to a gluelike consistency when wet, waiting to ensnare the unwary traveler in a pot of near-liquid cement. The glue hardens to form a plaster cast of your car when it dries. Even asphalt pavement is not always safe to combat this characteristic, for the Moab (Canyonlands) and Cortez airports have been known to almost sink from sight during exceptionally wet weather.

In spite of the stagnant depositional environment that the Mancos Shale probably represents, the formation contains a surprisingly rich fauna of fossil life. Among the types of fossil that are abundant are strange heavy-shelled clams, small coiled cephalopods, the tiny shells of the single-celled Foraminifera which are plentiful when the shale is broken down into a mush, and last but far from least, sharks' teeth. All these signify a marine environment of deposition, and an Upper Cretaceous age for the formation. Of the forms present, all but the clams either swam or floated near the surface, where the waters were probably far better oxygenated than in the deeper layers.

Regressive Traits

Unfortunately for geologists, the advance of the Upper Cretaceous sea was not regular and persistent. The result was that near-shore sand deposits formed along the shorelines as the sea first advanced and then retreated, forming within the Mancos Shale highly complex relationships of sandstone tongues that vary from place to place. Throughout most of the Colorado Plateau, where Cretaceous strata are preserved, the Mancos Shale's black slopes are overlain by two or more tan sandstones that formed during these fluctuations of sea level; their ages vary wildly and are never quite the same from one area to the next. Sandstone beds vary radically in their stratigraphic position within the black shale matrix and must be traced with great care along outcrops and between wells.

The best-known exposures of these sandstone tongues occur in the vicinity of Mesa Verde National Park in the northwestern corner of the San Juan basin. For this reason the post-Mancos beds were originally named the Mesa Verde Formation and more recently changed in rank to the Mesaverde Group. In that country the Mesaverde section consists of a lower massive sandstone cliff that caps the black shale slopes with a decidedly gradational contact. The lower cliff-forming unit forms the prominent buttress on the skyline at the entrance to Mesa Verde National Park and was named the Point

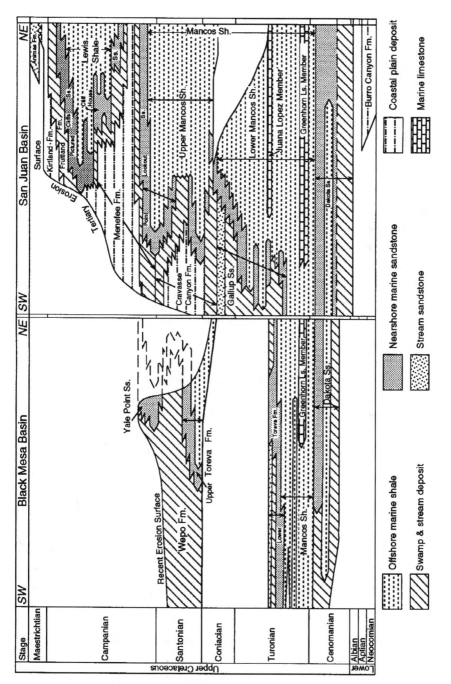

Generalized cross section comparing the names and ages of the very complex Cretaceous rocks of the Black Mesa and San Juan basins. The strange names along the left margin of the drawing are Stage names used for specific time units of the Cretaceous Period.

Lookout Sandstone for the topographic feature. It represents near-shore sand accumulations formed during a major retreat of the Upper Cretaceous sea.

The Point Lookout is overlain in turn by the Menefee Formation, which forms well-developed slopes above the lower cliffs. It is composed mainly of gray carbonaceous shale and sandstone, along with varying amounts of coal. This easily eroded formation was obviously deposited in environments that lay marginal to the sea, for the coal and related deposits were undoubtedly the result of sedimentation in marshes and lagoons that formed behind the beaches of the retreating sea. As the oceans changed their minds once again and began another march across the Four Corners region, the Menefee everglades were inundated with beach sand. The upper of the Mesaverde trio of formations, the Cliff House Sandstone, was deposited as a forewarning of another great marine flood.

Thus, in the type area, the Mesaverde Group consists of two thick, cliff-forming sandstone formations of near-shore character, separated by an intervening slope-forming formation of swampy affinities. The resulting topography delighted the early Indians in the region, the Anasazi, for the upper cliffs provided nearly perfect protection from their enemies. The middle shale eroded back so that the upper cliff was undercut to form broad, open alcoves that protected their villages from the elements and hosted localized springs. They took advantage of these geologic circumstances and built great villages under the Cliff House Sandstone, thus providing an excuse for the establishment of Mesa Verde National Park.

The tripartite group that caps Mesa Verde in southwestern Colorado and northwestern New Mexico appears to be a uniform series of formations without too many problems in the park area. When traced regionally from this handy focal point, however, it is readily seen that the sandstones occupy varying stratigraphic positions with regard to surrounding rock units. And worse than that, other sandstone bodies appear above and below the true Mesa Verde units, representing other fluctuations of sea level not recorded specifically in the park. The result is that a number of different formation names have been applied to the different sandstone tongues. These are far too numerous to describe in this general discussion, so it will suffice to refer all post-Mancos sandstone beds and their intervening shaly units to the Mesaverde Group.

Without worrying about the details of the records of the numerous transgressions and regressions of the Upper Cretaceous sea, the Mesaverde Group can be traced over considerable distances and into the hearts of all the larger basins on the Colorado Plateau. The Mesaverde sandstone beds form distinctive cliffs surrounding the entire San Juan basin immediately above the

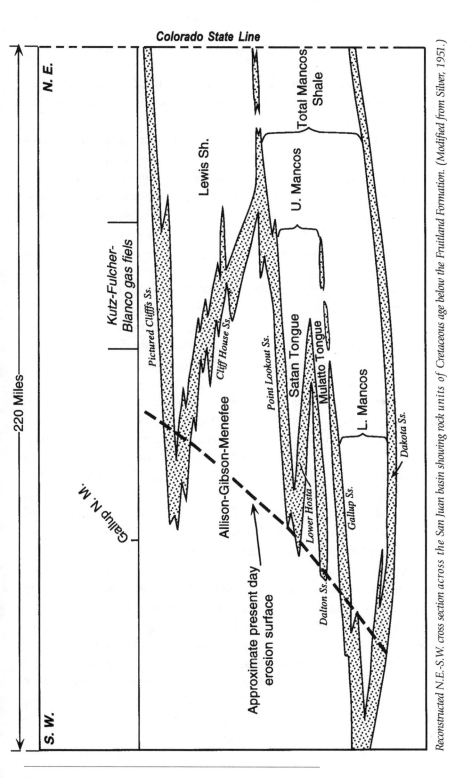

Reconstructed N.E.-S.W. cross section across the San Juan basin showing rock units of Cretaceous age below the Fruitland Formation. (Modified from Silver, 1951.)

gray Mancos Shale slopes. The prominent cliffs represent different sandstone wedges in different places, and many do not occur in Mesa Verde National Park. They all resulted from near-shore sedimentation of sands that migrated up and down the geologic section as the seas waxed and waned across the Cretaceous terrains.

The Mesaverde Group is represented by sandstone tongues in the Black Mesa basin east of Grand Canyon and south of Kayenta, Arizona, that are very similar in appearance to the Point Lookout-Menefee-Cliff House sequence. However, by a careful analysis of the fossils, the entire section here is found to be older than any part of the type Mesaverde Group; it corresponds in age to parts of the Mancos Shale in the park. Completely different formation names have rightfully been applied to these sandstone layers, although they form a very similar three-part group.

The underlying Mancos Shale gradually thins toward the southwest until the Mesaverde Group directly overlies the Dakota Sandstone in the vicinity of Show Low, Arizona. This probably resulted from the base of the Mesaverde Group being progressively older toward the southwest and the top of the Dakota becoming younger in the same direction, thus putting the squeeze on the Mancos Shale.

The Mesaverde Group also forms prominent cliffs above the Mancos slopes in the Book Cliffs of the northern Colorado Plateau between Price, Utah, and Grand Junction, Colorado. There, too, the section is almost identical in appearance to the type section, but the sandstone components are of different ages and affinities. Of course the individual formations cannot be traced continuously from one basin to the next, for intervening areas have been wiped clean of any residues of Cretaceous rocks across the uplifted regions. However, it is possible through paleontologic studies to determine that differences occur in the times of deposition in the various regions. It was through detailed studies of the relationships along the marvelous continuous exposures surrounding the San Juan basin and along the many tens of miles of outcrops in the Book Cliffs that the complexities of the Mesaverde Group were unraveled. If the rocks had been preserved in areas lying between the present basins it would be possible to work out the detailed correlations, but since recent erosion has robbed us of the once-intervening rocks it is impossible to make exact geologic connections between the widely spaced occurrences. For the moment it will have to suffice that the Mesaverde Group varies in age from one basin to the next, but that it everywhere represents similar conditions of the mobile shorelines.

The relationships seen in the northwestern corner of the Colorado Plateau reveal something about the origin of all of this clastic material that comprises the Mancos and Mesaverde formations. In the vicinity of Price, Utah, the entire Upper Cretaceous sequence becomes coarser and the sandstone begins to dominate over the Mancos-type shales. In working westward across the high plateaus that border the province, the Cretaceous rocks become thicker and rapidly coarser in grain size. They emerge on the west as a very thick conglomerate, which makes up the equivalent interval in west central Utah.

This information, along with the relationships already described in the Black Mesa basin, suggest that a highland formed in the country now known as western Utah and northwestern Arizona and shed large quantities of boulders and pebbles eastward toward the Colorado Plateau. As the sediments were transported increasing distances from the source region, the larger particles were left behind on continental lowlands bordering the uplift to form conglomeratic formations such as the Price River and North Horn in the high plateaus. The finer sand and mud was moved on eastward to become the Mancos and Mesaverde deposits.

Therefore, during any particular increment of Late Cretaceous time, the sediments can be seen to become finer in grain size toward the east and the open sea. The uplands formed to the west must have been of considerable magnitude, for thousands of cubic miles of sand were emitted toward the east to their final resting places in the great structural depressions of the Uinta, Black Mesa, and San Juan basins.

The Mesaverde sandstone strata are host to great quantities of natural gas in the San Juan basin. It is not surprising that they make good reservoirs, for the sandstones are usually highly porous and permeable—traits that are essential to the storing of any kind of fluids. What is interesting about the large reserves of gas is that they occur in the deeper parts of the basin. It has already been mentioned that petroleum is usually found in anticlinal or upfolded strata, for the hydrocarbons are less dense than the host waters found in almost all underground rocks. Consequently they rise to the top of beds and larger structures. In the San Juan basin, however, this is not the case and the gas lies in huge gas fields in the bottom of the basin, in contradiction to the simple laws of nature. The gas has been found immediately adjacent to the uparched borders of the basin, where it would seem that the gas should have surely escaped through the outcrops of the reservoir layers. Still the gas remains in regions of low structural position in this unnatural occurrence. What could cause this great enigma? Many explanations have been presented; most have been disproved to date.

In fact, some natural gas *does* leak out of the Mesaverde reservoir rocks, at least in a few isolated localities. There are unconfirmed stories of ranchers having ignited the gas seeps in the winter to keep their cattle warm. Known localities of these gas seeps are mostly along the Mesaverde hogbacks east of Durango, Colorado. This is not surprising given the many miles of exposures of the sandstone beds surrounding the San Juan basin.

Then in the late 1950s, a large consulting firm proposed that the gas was being drawn down in the central San Juan basin by natural processes. The study, which sold to several major oil companies, proposed that the gas was being sucked downward and trapped in place by osmotic pressure. The drawdown is due to Mesaverde reservoirs containing lower-salinity groundwater, being effected by the underlying Entrada Sandstone they believed would contain ground water with very high salinity. The intervening Mancos Shale would act like an osmotic membrane in the process. While drilling a deep hole in the heart of the basin, Shell Oil Company decided to test the hypothesis. They drill-stem tested the Entrada Sandstone to determine its water salinity. It was drinking water, not brine as proposed. Thus, a wild idea was put to rest.

The best explanation seems to be that the gas is being trapped from upward flow to the outcrops by hydrodynamic processes. Rain water naturally soaks into the up-turned edges of the exposed sandstone, and flows down-dip toward the basin interior. Ground water produced in this way tries to flow downward by gravity, but it encounters natural gas trying to flow up-dip toward the outcrop. Where the pressures from the two sources balance one another, an equilibrium is reached. The water can flow no deeper into the basin, and at least most of the gas is trapped. Some gas escapes in natural seeps, but most is trapped at depth.

Repeat Performance

The sea was not quite finished with the Colorado Plateau, for another thick formation of black shale was deposited by stagnant marine waters, at least in the vicinity of the San Juan basin. This unit, which is identical in appearance to the Mancos Shale, overlies the Mesaverde Group south of Durango, where it is called the Lewis Shale. Its marine faunas also attest to its more recent date of sedimentation, although conditions varied little from the previous Mancos regime. Records of this late Upper Cretaceous marine episode are not preserved in the other major basins of the Colorado Plateau, probably because of removal by later erosion in some cases and nondeposition in others. The presence of this formation is not unimportant, however, for it attains the significant thickness of at least 1,800 feet in the northern San Juan basin.

As the Lewis sea retreated, near-shore sands again covered the region of the San Juan basin now represented by a formation called the Pictured Cliffs Sandstone. The name was derived from an area near Farmington, New Mexico, where Indian petroglyphs decorate the cliffs formed by the erosion of this sandstone. The Pictured Cliffs deposits were undoubtedly formed in environments reminiscent of the Mesaverde Group, representing the last withdrawal of the sea from the province. Although the formation resulted from the final recession of a once great marine domain, the Pictured Cliffs is not without considerable importance, for it is the host of vast accumulations of gas in the deeper northern portions of the San Juan basin. In this respect it also mimics the Mesa Verde habits already established, for the gas again occurs in the anomalous downfolded fields in the same region as the Mesa Verde occurrences. Why does the very mobile gas remain in the basin and not leak out in nearby exposures? Gas producers care little about the why. Instead of worrying about the academic questions they spend their time praising Vishnu for such geologic incongruities.

Several formations mark the return of continental conditions in latest Cretaceous time. In the San Juan basin the Fruitland, Kirtland, and McDermott formations represent thick deposits of fluvial and related nonmarine sediments that accumulated in the last minutes of the Cretaceous Period. The source of the sediments was the initial uplift and volcanic eminences of the great rise of the San Juan Dome to the north. No latest-Cretaceous nonmarine counterpart is known in the Black Mesa basin, but the Kaiparowits Formation in the Kaiparowits basin and the North Horn Formation in the Uinta basin are probably correlative in time and mode of sedimentation.

Summary

Although the Cretaceous Period represents only some 70 million years of earth history, it was prolific as a time of marine sedimentation. Thick preserved deposits are found in the San Juan basin of northwest New Mexico and southwest Colorado, where more than 6,800 feet of sediments, mostly marine black shale, can be measured in the immediate vicinity of Durango, Colorado. All these formations are Cretaceous in age; some 99% of the section was deposited in Late Cretaceous time alone. The period is also of prime economic importance, as considerable quantities of oil are presently being produced from fossil offshore sand bars in the Cretaceous section near Farmington, New Mexico, and almost the entire northern half of the San Juan basin is productive of natural gas.

Dinosaurs, which dominated the land areas of the earth during the Mesozoic Era, were not to be found in the Cretaceous strata of the Colorado Plateau until the sea retreated in the closing moments of the time interval. Even then, they did not become particularly numerous. Some omnipotent series of events marked the close of the Mesozoic, however, for the giants of the reptile world, along with several important forms of marine life such as the ammonites, gradually died out, not a momentary end.

The reasons why such seemingly healthy and abundant forms of life should become extinct by the close of the Cretaceous is one of the great mysteries of historical geology. Perhaps the climate changed radically. Perhaps the desire to reproduce was terminated by temperature changes or psychological inhibitions. Maybe the vulnerable species simply evolved themselves into oblivion. Both dinosaurs and ammonites became extremely ornate with seemingly useless physiological characteristics as the geologic period waned.

Perhaps their supply of food was suddenly cut off by some catastrophe that is only visible to the physicist's crystal ball. It is fashionable today to blame it all on a huge meteorite that fell to earth in the Caribbean exactly 65 million years ago—the end of the Cretaceous. Yet the species were becoming fewer and more weird even as the period was gradually coming to a close. The meteoritic dust cloud, that supposedly killed off the dinosaur's food supply, was not as affective as postulated. The grasses and flowering plants that first appeared in Cretaceous time flourished into the Tertiary Period! Tiny mammals survived the supposed catastrophe and became dominant during the Tertiary! Most of the mollusks went on to happier fruition with the turning of the calendar! Extinctions were slow and gradual, not momentary. For now the physicists and geophysicists rule. Thinking geologists who look at the overall biotic picture should know better. Again: "One finds what one is looking for!"

At any rate, when the last dinosaur stumbled and fell headlong into the bog, the Cretaceous Period, by definition, came to a close.

CHAPTER NINE

OROGENY MOLDS THE LAND

TERTIARY TIME

As Cretaceous time drew to a close the lands were becoming noticeably higher and the seas were withdrawing toward the basins in which they now reside. An episode of warping and buckling of the earth's surface in the western United States was beginning that was unprecedented since Precambrian time and would permanently mar the face of the Colorado Plateau. Indeed, mountain building was already well under way in the Great Basin country of Nevada and western Utah, having begun in Late Jurassic time in the Sierra Nevada and spread gradually eastward toward the Colorado Plateau as Cretaceous time progressed. The folding, normal faulting, thrust faulting, and volcanism of the western region effectively shut out the sea from the Cordilleran seaway for all time. The creeping catastrophe didn't reach the western borders of the Colorado Plateau until near the close of the Cretaceous, when previously deposited strata were tilted, folded and subsequently eroded to form angular unconformities in the closing millennia of the period. The first effects of the advancing wave of mountain-building forces was felt in the so-called High Plateaus, which lie along the western margin of the province.

Times of concentrated folding and faulting of the earth's crust of considerable magnitude are known in the profession as "orogenies." In the early days, when geology was in its infancy, scientists generally believed that such times of mountain building were of worldwide extent and that these highly significant events were essentially instantaneous catastrophes. Consequently, orogenies were used as the valid basis for subdividing the geologic column into the various geologic periods. In other words, orogenies were thought to be worldwide punctuation marks in the geologic history book. Wherever a major unconformity, or break in the record, was encountered, it was taken as an orogeny that happened in a very brief time at the close of a geologic period.

It is now known, largely as a result of studies of the Late Cretaceous and Early Tertiary orogeny in the western United States, that mountain-building processes are slow and may affect relatively small regions for time intervals that encompass parts of two or more consecutive geologic periods. In the case of the western orogeny, folding and faulting began in Late Jurassic time in the western Great Basin and was for a long time segregated as the "Nevadan orogeny." Most of the mountain-building activity in the Colorado Plateau and Rocky Mountains occurred in latest Cretaceous and Early Tertiary times and was set aside as the "Laramide orogeny."

It is now seen to be altogether possible that the entire sequence of events was interrelated, and that the forces that contorted the strata moved in a slow wave that migrated eastward with time. Mountain-building forces affected the various local areas in turn from west to east. At the very least, it is now a well-known fact that the "Laramide orogeny" did not occur in a brief span of time at the close of the Cretaceous. Instead it was actively disturbing the status quo from late in Cretaceous time on into perhaps even middle Tertiary times. Regardless of when one wants to start and stop the various orogenies, the folding and faulting and volcanism came to a climax in the western United States well into Tertiary time. Thus ended a prolonged siege of destruction that should be given a more general name, such as "Cordilleran orogeny."

The dating for the various local events is usually accomplished by the relative times of production of unconformities and the sedimentary accumulations that ensue from the erosion of the uplifted elements. For example, if the Upper Cretaceous Mesaverde Group were tilted or folded and eroded prior to the deposition of the Middle Tertiary Wasatch Shale, the time of folding would be determined as post-Mesaverde and pre-Wasatch. As the dates for the deposition of those two formations is known, it could be further stated as post-Upper Cretaceous and pre-Middle Tertiary.

As often happens, uplifted regions undergo erosion in the process of their tectonic development, and the sediments produced from the erosion must go somewhere to form other deposits. If these new formations contain diagnostic fossils it is possible to date the time of folding or faulting more accurately by dating the resultant sedimentary rocks. The time of uplift of the above hypothetical example might be thus determined to be Early Tertiary in age to narrow the time of the event to a very slim margin.

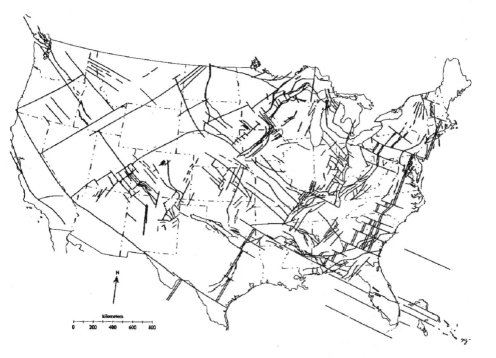

Map of the conterminous United States showing the distribution of major basement fault systems believed to have been present in Late Precambrian time. The fault systems are considered to be wrench fault lineaments oriented both to the northeast and northwest in a consistent pattern. All continents contain similar basement patterns. The northeasterly faults moved consistently toward the left (left-lateral displacement), while the northwesterly fault zones show movement toward the right (right-lateral displacement). This pattern indicates that strong compressive forces in the earth were oriented north-south in Precambrian time. Note the congested fault pattern in the Four Corners area. From Baars and others, 1995.

Further dating, by determining the rates of radioactive decay of certain elements such as uranium, thorium, rubidium, and potassium and the ratio between the parent element and the newly formed daughter elements found in the rock, can be utilized to obtain ages in years. These techniques are usually employed only to date igneous rocks, for that is where most of the radioactive elements occur. Consequently, the time of emplacement of the various kinds of igneous bodies can be dated with some degree of accuracy, if the proper minerals are present and the necessary money for the determination is forthcoming.

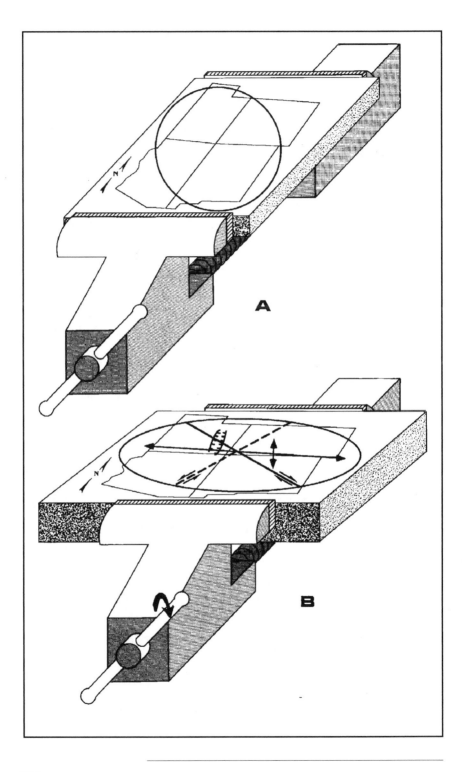

Roots

Recent developments in evaluation of the basic geologic framework of the Colorado Plateau have shed new light on the origins of the prominent geologic structures that have heretofore been elusive. It has long been in vogue to blame the prominent uplifts and basins and their related smaller folds and faults on the Laramide orogeny. However, to do so creates many unresolved problems, because the obvious features are not oriented properly to conform to known directions of forces active in the crust of the earth at that time. We have already seen that the shortening of the crust resulting from compression began in the west in Jurassic time and progressed eastward through Cretaceous and Early Tertiary time. This resulted from a strong driving mechanism or force directed from west to east. It would cause geologic structures that are not present on the Colorado Plateau or Southern Rocky Mountains to form in the stress field. Where, then, did we go wrong?

Geologists have long been bewildered at the last significant structural event, the Laramide orogeny. This century of confused interpretation invariably overlooked the obvious fact that several episodes of folding and faulting preceded Laramide time, leaving a legacy of structural features to be modified by later Laramide deformation. When this legacy is carefully examined, it becomes apparent that all the structural features at the surface of the Colorado Plateau and adjacent Rocky Mountains have been in their modern positions and orientations for millions of years prior to the Laramide events. The basic structural fabric was present all the time, only to be distorted by Laramide forces. But, you say, what difference does this make?

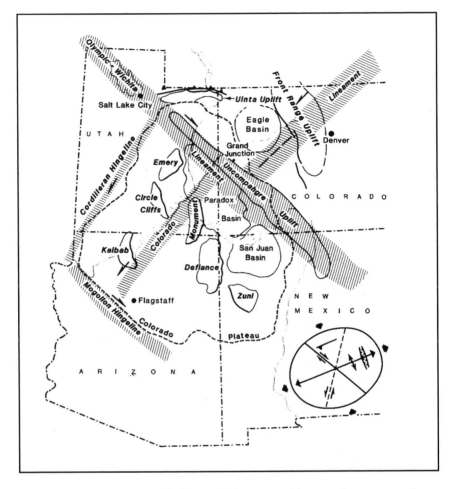

Index map showing relationship of prominent surface structural features to basement wrench fault systems, shown generalized as shaded lineaments. Strain ellipsoid in lower right corner indicates expected orientations of shears, faults and folds in the Precambrian stress field. (Baars and Stevenson, 1981)

We have previously seen that swarms of wrench faults of considerable magnitude segmented the western United States into orthogonal blocks some time around the summer of 1.7 billion years ago (map, p.49). Two of the fault systems transected the eastern Colorado Plateau, intersecting in the vicinity of Moab, Utah. The northwest-southeast oriented lineament, the Olympic-Wichita, dominates the Paradox basin, and where it crosses the northeast-trending Colorado lineament, the latter trend is offset to the right (right lateral displacement). Conversely, the Colorado lineament is seen to offset the northwesterly set to the left (left lateral displacement). This is shown with arrows on the maps (on pages 49 and 180). That the two fault systems displace each other indicates that movement on both lineaments was more or less simultaneous and genetically very closely related, forming what is called a "conjugate set."

The egg-shaped inserts on both maps are called strain ellipsoids, and result from decades of study of what is expected to happen when rocks are subjected to various stresses. The X- shaped lines within the "egg" show how the rock should fracture if the forces are directed as shown by the external arrows; the "egg" is oriented on the maps to coincide with the basement lineaments on the eastern Colorado Plateau. One can see, then, that if everything is properly interpreted on the maps, the primary direction of compression, or squeeze, was from the north and south. If the force field were directed from the west, as in Laramide time, the sense of displacement of the wrench faults would be reversed by necessity. This, however, is not the case, as revealed by observation in the field. The Precambrian compressional force was from the north, while the Laramide force came some ninety degrees differently, and from the west.

The strain ellipsoids (eggs) give further insight into the structural configuration of smaller structures in the province. When the stress was directed from the north in Precambrian time, one would expect normal faults that result in extension, or stretching, of the crust to be oriented in a north-south direction and the stretching in a west-to-east direction. The heavy black lines bounding the eastern limits of the major surface uplifts of the Plateau, the Kaibab, Circle Cliffs, Emery, Monument and Defiance uplifts on the map (on page 180), are deep-seated (basement) normal faults over which the younger rocks were later draped to form the large monoclines that are so typical of the province. The large faults that bound the major uplifted blocks of the Colorado Rockies are also normal faults that date from Precambrian time.

The Uinta Mountains of northeastern Utah have plagued structural geologists for over a hundred years. The east-west trending mountain range is a very large anticline whose flanks are bounded by thrust faults, resulting from compressional forces from the north that shortened the crust. Such a situation is the expected one in the Precambrian scenario, but impossible if studied from the Laramide point of view. Angular unconformities in the western Uinta Mountains indicate that the east-trending fold was present in its modern configuration by Cambrian time, thus predating the Laramide by at least a half billion years. Thus the enigma of the Uinta Mountains is laid to rest and the peculiar structures of the Colorado Plateau made simple.

The Laramide orogeny affected the Colorado Plateau province by overturning the existing basement structures toward the east, and enhanced the structures to their present grandeur.

The Laramide Orogeny

As Tertiary time gradually became a reality, the strata laid down in Cretaceous and all earlier times were affected in one way or another on the Colorado Plateau. At least in the broad sense, the folding and faulting that ensued followed predetermined patterns that can in many cases be traced back in time even into the Precambrian Era. Examples are omnipresent. The large-scale faulting previously described affecting Precambrian and Paleozoic rocks in the central San Juan Mountains was rejuvenated in Laramide time (Upper Cretaceous to Early Tertiary). The major uplifts known as the Defiance and Zuni were formed in their present configuration in Laramide time. They were also high during most of Paleozoic time and perhaps also in Precambrian time. Other major uplifts such as the Kaibab, Monument, and San Rafael Swell have had intermittent periods of uplift during most of Paleozoic and perhaps Precambrian times, even though they took on their present form during the Laramide orogeny. The basins also were involved. The San Juan, Black Mesa, and Paradox basins were all actively subsiding during various times in the Paleozoic and Precambrian Eras, and were further depressed by the Laramide forces into the structurally low features we see today.

So it seems that nothing very new happened, but instead the previously existing structural weaknesses in the crust were reactivated and reemphasized concurrently during the Laramide mountain-building episode. The underlying causes for the localization of the particular structural elements can usually be traced to the tectonic fabric of the basement rocks. At any rate, virtually every fold and fault on the Colorado Plateau was reactivated during Upper Cretaceous and Early Tertiary times, regardless of the underlying causes. Almost without exception, every structure seen today can be attributed to Laramide forces, whether or not it was also present at an earlier date in history. Therefore, the Laramide orogeny was the single most prominent event in the molding of the geologic structure of the Colorado Plateau as it is seen today.

As the uplifts actively rose, the basins actively subsided in Laramide time. The result was that the positive areas were attacked by erosion to form vast quantities of sediments, which were in turn transported to and deposited in

An abandoned silver mining operation in La Plata Canyon, west of Durango, Colorado, in the La Plata Mountains.

the sinking basins. The San Juan basin is perhaps the best example of this relationship. The faulting and folding in the San Juan Mountains began ex- aggerating the uplift in Late Cretaceous and Early Tertiary times. The uplift produced coarse sediments that were transported down the flanks of the dome toward the south to be deposited in the deeper parts of the San Juan basin, where the streams flowed onto the flat country and lost their carrying potential. The resulting accumulations of thick fluvial sediments can be seen in the heart of the basin, especially along U.S. Highway 550 between Durango, Colorado, and Aztec, New Mexico, and then along State Highway 44 to Cuba, New Mexico. While it is probable that most of the Tertiary deposits were derived from the erosion of the San Juan dome, some of the clastic debris probably also came from the actively uplifting Defiance, Zuni, and Nacimiento positive features, which surround the basin on the other three sides. These Early Tertiary stream deposits are usually included in the McDermott, Animas and San Jose formations.

Another area that received thick continental deposits derived from the Early Tertiary tectonism was the Uinta basin, in the northern Colorado Plateau. There thick stream and lake deposits known as the Wasatch and Green River formations, in ascending order, were derived in part from the erosion of the rising Uinta Mountains to the north. As subsidence on either flank of the Uinta Mountain arch continued, the Green River lake filled the surrounding structural depression and thick lake deposits resulted. Uplift of the entire Colorado Plateau began with tilting of the province toward the north, and consequent erosion and stream transport supplied sediments to the vast settling basin of the lake. So uplifted regions provided the sediment supply and down warping basins trapped and preserved them, entirely in continental environs.

Other accumulations of Tertiary sediments are known on the Colorado Plateau, but they are much more local in occurrence usually of a Middle to Late Tertiary age. Examples of these younger formations are the Chuska and Bidahochi, of northeastern Arizona.

The Fiery Days

As often happens, the folding and faulting of the Colorado Plateau strata in Laramide time were accompanied by an outbreak of volcanic activity. The first of the fiery eminences is recorded by the McDermott Formation of latest Cretaceous age immediately south of Durango, Colorado, which is composed of sand and gravel derived from igneous terrains. The Cretaceous Period was on its last leg, however, and the bulk of the volcanism occurred in the early half of the Tertiary Period. With the intensification of the structural deformation came an increase in mobility of the molten rocks, and large-scale intrusion and extrusion of the "instant rock" resulted.

When a large body of molten material, or magma as it is called, intrudes an area of the earth's crust, it does so by forcing its way into the weaknesses of the crustal structure, along faults and fractures in many cases. In other instances the intrusion simply melts the surrounding rock and assimilates the material into the caldron. When the magma cannot "elbow" its way to the earth's surface, it cools very slowly beneath the crust, and rocks with large crystals develop. Granite is a well-known rock of the type, and the coarsely crystaline textures that ensue are often called granitic. The granitic bodies are found in several forms. Those that squeezed into place along fractures make tabular-shaped bodies that cross-cut the layered rocks and are called "dikes." Others force their way along preexisting bedding planes to form concordant

Shiprock, New Mexico, as viewed from the air. The prominent landmark rises some 1800 feet above the desert floor, which is here composed of the Late Cretaceous Mancos Shale. It is a neck or plug of a Tertiary volcano that has since been eroded away except for this more-resistant internal remnant. The force of the explosive eruptions created radiating fractures that were infilled with igneous material to form the three almost symmetrically arranged dikes seen radiating from the plug. Shiprock remained unclimbed until the early 1940s, but was conquered by hundreds of climbers before the Navajo Indians put a moratorium on rock climbing on the Reservation.

tabular intrusive bodies that are distinguished as "sills." In some cases on the Colorado Plateau the intrusive magmas were injected through dike systems from below only to spread out laterally along the bedding planes as sills. When the supply of molten materials continued to arrive at the sill region, the upper layers of overlying strata were bowed up to form mushroom-shaped bodies of igneous rock within the earth's crust, which are called "laccoliths." If larger regions that can be measured in numerous square miles of granitic rock result from massive intrusions they are known as stocks and batholiths to distinguish them from their smaller relatives.

If on the other hand the igneous melt is able to reach the surface and flow out onto the ground, a lava flow and extrusive rock result. When this happens the magmatic material cools very rapidly to form a very finely crystalline rock texture, and even natural glass may be formed. The flows originate in a volcano, which in turn is fed by the magma at depth. The magma may still cool and harden into an intrusive rock. When the molten caldron is highly charged with gas, the eruptions of the volcano may be sporadic and cause explosive emissions that strew volcanic ash and cinders and other debris over vast areas. Very little lava results from these gaseous eruptions, but sediment-like volcanic materials may accumulate in important deposits. In some cases the original volcano may be completely ravished by erosion, wiped forever from the scene. The necks or pipes that fed the volcano are often exposed to view by such a process. If the plug is more resistant to erosion than the surrounding rock layers, such a neck or plug may form prominent topographic projections at the earth's surface.

Volcanic activity apparently began in the San Juan Mountains of southwestern Colorado in latest Cretaceous time, but didn't really make a spectacle of itself until early in Tertiary time. The deposits formed were largely of the ash and cinder variety, but lava flows are occasionally interspersed with tuffaceous (ash and cinder) material. The volcanic centers were several in number. One of the largest forms a circular pipe of lava that is now downfaulted into a caldera immediately north of Silverton. The stream valleys of the Animas and Mineral creeks, which extend toward the northeast and northwest from Silverton respectively, follow the margins of the ancient volcanic vent. The caldera extends for several miles to the north, with its opposite margin passing north of Red Mountain Pass.

Other major sources for the volcanic material are known in the region, with well-developed sister calderas present near Lake City and Creede, Colorado. These volcanos emitted great quantities of extrusive materials of various

Agathla Peak, in the Navajo language "A place to shear the sheep," or El Capitan as named by Kit Carson, near Kayenta, Arizona. A Neck, or plug, of an explosive volcano of Tertiary age.

types. They eventually blanketed the central and eastern San Juan Mountains and adjacent regions with several thousand feet of tuffaceous and flow rocks. The gray layers that cap the mountains in the vicinity of Silverton and Ouray are volcanic in origin, and the entire mountainous region northeast and east of Silverton is eroded from these extrusive deposits.

The volcanos continued to belch fire and ash throughout most of the early half of the Tertiary Period (through Miocene time), creating an inferno on the crest of the ancient faulted dome of the San Juan range. The original extent of the inferno was probably more widespread than it now appears, for erosional remnants of similar material of the same age are common through central Colorado, especially in the Gunnison, Crested Butte, and Salida regions.

Volcanic lavas of Early and Middle Tertiary age are also widespread across several areas surrounding the margins of the Colorado Plateau. One of the larger of the lava fields lies to the southwest of Albuquerque and west of Socorro, New Mexico. This region, known as the Datil volcanic field, is several thousand square miles in geographic extent and untold hundreds or even thousands of feet in thickness. The volcanic flows extend westward into the White Mountains of Arizona almost to Globe and bury vast regions between Morenci and Fort Apache under the once-viscous sea of fire.

Tertiary lavas are also found sporadically across central Arizona along the southwestern margin of the Colorado Plateau, occurring mainly south of Interstate Highway 40, and adjacent to the province on the west in the vicinity of Lake Mead. Still another vast region buried under Tertiary volcanic flows lies along the northwestern margin of the plateau. It extends northward from about St. George to Marysvale and Ritchfield, and eastward almost to Bryce Canyon and Capitol Reef national parks. The latter volcanic field provides the cap rocks for most of the so-called High Plateaus that border the Colorado Plateau province on the west. Thus, the province was almost completely encircled with a tempest of fiery emanations during the early Tertiary stages of the Laramide orogeny, with the only escape route being northward across the Uinta country into Wyoming.

Just as the Colorado Plateau was ringed with volcanic activity, the San Juan and Black Mesa basins of the southern half of the province were surrounded by small eruptive vents of the explosive variety. The explosive igneous pipes are known as "diatremes." They are seen today as the plugs of Tertiary volcanos, standing in bold topographic relief above the surrounding plains of the soft Mancos Shale or other closely related formations. Several of these prominent plugs are landmarks well known to both modern man and the early Indian inhabitants, who commemorated them in legend. Among these are such familiar peaks as Shiprock, Cabezon, and Agathla, to say nothing of the several hundred lesser buttes and spires that dot the landscape around the basins.

It is likely that these smaller vents were localized as a result of tensional fracture systems that develop where the rocks begin to drape over the basin margin, stretching the earth's crust to form passages for the eruptive materials to follow. The explosive vents not only discharged volcanic materials, but their violent eruptions also tore loose large fragments of the country rock and flung them out at random. The debris that fell back into the open vent was composed of both volcanic and sedimentary rock fragments, and when the hot volcanic material cooled, a plug was formed that contained particles of all rock types through which the vent traversed. Upon close inspection of many of the exposed plugs, the actual plugging material is clearly seen to contain fragments, as large as automobiles, of red sandstone and shale and even an occasional granite fragment scattered throughout the rock mass. Clearly, this was no place to loiter without an umbrella in Early Tertiary times.

Meanwhile, in the depths of the earth's crust, other bodies of molten igneous magma were brewing, but didn't make their way to the surface. The liquid rock was able to work its way upward to some extent by forcing open already present fractures and faults and forcing its way into the available openings. When the broth encountered available bedding planes between rock layers, it spread out laterally to form sills. The sills expanded in thickness to the point of forming very fattened sills that bowed up the overlying country rock to form "laccoliths."

The laccolithic complexes developed in a number of localities across the Colorado Plateau in more or less random patterns. The reasons for the particular chosen localities for intrusion is not evident in most cases. Some geologists have tried to show that they occur in linear patterns that parallel the ancient fault system underneath the Paradox basin. Since there are seldom more than two laccolithic ranges that line up, it is difficult to make a convincing argument, although the idea is probably sound.

One of the intrusive ranges, the La Sal Mountains just southeast of Moab, Utah, took advantage of a previously existing salt-intruded anticline for its entry path. The primary igneous bodies of the range lie directly between salt structures, and parallel their trend exactly. Other ranges, such as the La Platas, the Utes, the Abajos, the Carrizos, and the Henrys, show no convincing patterns or obvious controls for their locations.

All these laccolithic ranges are prominent landmarks on the Colorado Plateau and can be seen for a hundred miles or more; usually two or more of the isolated ranges can be seen from any one point on the plateau. Although recent geologic studies have shown that most of the igneous bodies are not truly laccolithic in shape, when viewed at close range they appear to consist of local and very thickened sills forming shoulders and buttresses along the margins of the ranges.

All this igneous activity was not merely pyrotechnic display, for something of economic good was developing as a by-product. In the San Juan Mountains of southwestern Colorado, the igneous events were accompanied by mineralizing juices that permeated the igneous rocks themselves as well as some of the adjacent country rock. The rich fluids were apparently derived from the parent magma and penetrated upward through the crust along with the molten materials. The ores were emplaced by reaction and replacement of the igneous and enclosing sedimentary rocks with the mineralizing fluids in the dying stages of the fiery catastrophe to form rich metallic ore deposits. The resulting minerals include such common species as pyrite, chalcopyrite, galena, and sphalerite, plus dozens of lesser-known varieties. The primary metals that are derived from these minerals are copper, lead, zinc, silver, and a little gold. Enough gold was found to excite the prospectors at about the turn of the century and to keep such mining towns as Silverton and Telluride prosperous for half a century. The relative economic depression in the mining industry has shut down most of the mining activities in the San Juan Mountains.

The mineralization process was not restricted to the igneous events of the San Juan Mountains, but also was active in and adjacent to some of the laccolithic ranges. Base metals were not the primary deposits, and mining of their ores is quite limited in the desert ranges. However, it seems very probable that the juices that supplied the region with prolific quantities of uranium minerals came from the same source as the laccolithic igneous materials. For one thing, the mineralization of the Uravan Mineral Belt rings the southern and eastern fringes of the La Sal Mountain intrusives, a relationship that appears to be more than coincidental. Then, too, most of the uranium ores that have been dated by radioactive dating techniques reveal an age of mineral emplacement that approximately coincides with the age of the intrusive igneous activity, or about Middle Tertiary. Furthermore, the mineralizing fluids had to come from someplace, and the extensive igneous activity provided a ready mechanism.

Geophysics or Metaphysics?

Over the past couple of decades, it has become quite fashionable in the geological community to relate all structural episodes such as described for the Colorado Plateau to "plate tectonics." The general concept of plate tectonics is based on the assumption that vast blocks of the earth's crust are easily detached from the core of the earth along a shear zone at or near the mantle. Such blocks, or "plates," can wander about the surface of the globe rather promiscuously, being motivated by convection cells and their currents in the outer core region of the earth. Naturally, the mobile plates must collide when conflicts arise. Old plate materials must be reassimilated into the depths of the earth along "subduction zones" as new crust is formed along the midoceanic ridges where magmas rise to the surface.

Consequently, the surface of the earth is something like a bumper car ride at an amusement park, with major fender benders occurring where plates collide and new cars always being added as old ones are melted down. The continental plates (red cars) are lighter and more buoyant on the mantle, and consequently ride higher and last longer than the oceanic plates (blue cars). Blue cars are occasionally overridden by the red cars and have to be periodically recycled by melting down and rebuilding. The red cars were all manufactured at one or two factories, Gondwanaland and/or Pangea, and have since been distributed throughout the globe on convection conveyor belts to combat and override the blue cars as necessary to keep the system operative.

The concept of plate tectonics may be likened to a new religion. Since hard facts are lacking, if one is not a "believer" one is considered an "atheist" with regard to the many theories and interpretations of the "clergy": The oceanographers and geophysicists. Many of the concepts are plausible and exciting, and sometimes they fit the hard geologic facts. Many times, however, they are contradictory and totally incongruent with known geologic facts, at which time the facts are ignored. With enough "faith," every known earth event is compatible with the religion, especially with respect to oceanography. On land, however, where outcrops and fossils abound, it is often extremely difficult to be a "follower." The entire doctrine may in time be proven true, it may be completely disproven by geologists, or a compromise may be reached. I prefer to think the last possibility is likely.

Originally, presumably in the earlier Precambrian Era, all of the continental plates were one or perhaps two continents. As the earth began to cool and the outer crust solidified, great orthogonal fracture patterns developed throughout the crust. This seems to be a plausible explanation for the basement rift zones that occur in the eastern Colorado Plateau and, indeed, universally. As convection cells developed beneath the crust (if that is even possible), the "plates" began to gyrate, and seaways opened and closed randomly within the continent and ocean floors. It is extremely difficult to bring reasonable organization to the plate tectonic confusion that prevailed, as physical data are scarce at best. However, by Triassic or at least Jurassic times, the continents began their great migrations away from the homeland. North America and South America are two distinct continental plates that acted independently. Both generally wandered westward from the African and European sister plates. The North and South American plates had already been in conflict during the late Paleozoic, according to a recent interpretation that attributes the Ancestral Rockies orogeny to a plate collision between the two continents. We have seen, however, that the resulting structures developed along preexisting basement fractures. Nevertheless the two major plates drifted lazily westward in early to mid-Mesozoic times, setting up waves of crustal forces from west to east across the leading margins of the plates as they overrode the Pacific oceanic plates.

At this point, the story agrees with the known structural history as previously outlined, and handily explains the Nevadan-Laramide structural events. Whatever the underlying cause, intense structural deformation begin in the Sierra Nevada and progressed eastward across the Great Basin and into the Colorado Plateau in this time framework, providing some of the best geologic evidence for plate tectonics.

The basic plan is, however, beset with problems, as the basement rift zones will attest. Northeast and northwest trending basement fractures are well documented throughout the earth. All the plate movements now proposed require rotation of the continental plates to attain the known present-day product. That being the case, the basement rifts should be oriented differently on different continents, but they are not. Furthermore, three of the known basement lineaments in western North America, the Olympic-Wichita, the Walker Lane-Texas, and Mendocino lineaments, all cross from the North American plate onto the Pacific plates. This should be impossible if the Pacific plate has been undergoing subduction under the continental margin since the Triassic. Even coincidence would not permit three widely separated and differently oriented lineaments to coincide after all that plate wandering. And the present-day San Andreas rift zone in southern California parallels its Precambrian ancestors to the east, both geographically and in sense of relative displacement. Is that a coincidence, or is the San Andreas system just another rejuvenated basement structure?

To reiterate a thought expressed above, regardless of the underlying mechanism, the Nevadan-Laramide orogeny served only to disfigure a structural fabric that was well established by Late Precambrian time.

It would require another book to argue fully the pros and cons of plate tectonics theory. It is obvious at this point that I have not been totally converted to the religion. That is a matter for individual preference. You are free to believe as you wish, but please, don't send missionaries!

Bodily Uplift

The folding and faulting activity of the Early to Middle Tertiary Period molded the Colorado Plateau into the structural configuration we see at the surface today. The major uplifts and basins were formed, as well as the small local anticlines and normal faults. The accompanying igneous activity buried much of the San Juan Mountains and areas fringing the province under hundreds of feet of lavas and tuffs, and simultaneously produced the isolated laccolithic desert ranges that dot the horizon. Although these processes are generally thought to be destructive, the folding and faulting provided the structures necessary for the entrapment of oil and gas, and the igneous activity supplied the province with a wealth of minerals. In late Tertiary time, however, most of this potential wealth was buried deep within the earth's crust, beyond the reach of any of man's modern tools of exploration.

The processes of erosion were not dormant during the creation of all this havoc, however, and the uplifts and volcanoes were subjected to the destructive forces of wind and rain, running water, and freezing temperatures from the beginning. The effects were probably negligible while the land lay very near sea level, during the early half of the period, but a general rise of the continental interior began in Late Tertiary time that greatly accentuated the erosive mechanism. Not only did it affect the Colorado Plateau, but the entire Rocky Mountain region and all adjacent areas were bodily uplifted to the tune of several thousand feet. The result was that elevations changed fairly abruptly from near sea level to about the 4000–6000 foot elevations we see today, and erosion was accelerated to begin carving away the excess strata with unprecedented fury.

CHAPTER TEN

EROSION SCULPTS THE LAND

PLEISTOCENE TIME

The principal drainage patterns that score the Colorado Plateau today were undoubtedly already established in Late Tertiary time. The denudation of the larger structures progressed from their very inception into Late Tertiary and Pleistocene times. The excess strata were carved away to produce the colorful geological fairyland of deep canyons, high plateaus, and hogbacks we know today. Although the results of the natural labors are obvious and the mechanisms are today well understood, the details of the history of development of the present erosional patterns are among the least understood aspects of the geology of the Colorado Plateau. Many enigmas remain. The primary problem in a nutshell is: How can a river carve its path directly into the very crest of a major uplifted area and come out the victor?

The Colorado Plateau is geomorphologically unique in that river after river crosses the larger uplifts, showing no regard for convention or the laws of nature. Instead of flowing around the high structural features, as any respectable river would do, the rivers of the plateau flow directly into the uplifts. They carve impressive channels across the highest parts, and emerge on the opposite flank as if this were the easiest thing to do. The great canyons of the west result: The Gates of Lodore and Split Mountain canyons across the Uinta Mountains uplift; Cataract Canyon across the northwest flank of the Monument Upwarp; the spectacular canyons of the San Juan River across the very navel of the Monument Upwarp; the Colorado, Dolores, and San Miguel rivers cutting across the great salt-intruded anticlines at right angles; and, far from the least impressive, the Grand Canyon of the Colorado River crossing the southern nose of the Kaibab uplift. What enables these great rivers, but

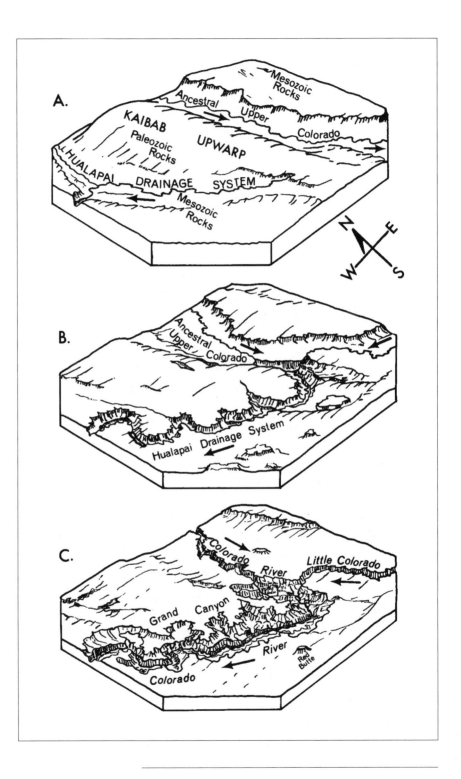

A.

Mesozoic
Rocks

Ancestral
Upper
Colorado

KAIBAB
Paleozoic
Rocks

UPWARP

HUALAPAI DRAINAGE SYSTEM

Mesozoic
Rocks

N
E
W
S

B.

Ancestral
Upper Colorado

Hualapai Drainage System

C.

Colorado River

Little Colorado

Grand Canyon

River

Red
Butte

Colorado

THE COLORADO PLATEAU

FACING PAGE: *Structural superpositions of the Hualapai drainage system across the kaibab upwarp, and capture of the ancestral upper Colorado River, in post-early Pliocene time.*

A. *Structural superposition by headward erosion of the Hualapai drainage system;*

B. *Capture and diversion of the ancestral upper Colorado;
reversal of direction of stream flow in the Little Colorado drainage basin;*

C. *Cutting of eastern Grand Canyon and the canyon of the Little Colorado River.
(After Strahler, 1948.)*

————————————————————————————

not others, to conquer great uplifts on a wholesale basis?

Early geologists to visit the plateau country, such as Powell and Dutton, realized that everything was quite unusual as they traced the courses of the major rivers by boat and on horseback. They reasoned that in order to pro-duce such unusual canyons through the uplifts, the rivers must have been flowing across the countryside before there was any structure evident at the surface. They felt that the solution must be that the uplifts formed slowly in the path of the particular river, and that the rivers were able to keep pace with the arching of the rocks by down-cutting at the precise rate at which the structures were rising. In other words, they believed that the rivers were "an-tecedent" to the structures and maintained their courses by increased rates of erosion even as the uplifts grew in their paths.

This hypothesis would require that the rivers be already flowing in their present courses in Early Tertiary time, when the uplifts and basins were form-ing. More-recent work has shown, however, that the drainage of Early Ter-tiary time was primarily from the uplifts into the basins, where great thick-nesses of sedimentary strata were deposited as mute evidence. The present courses of the major rivers are not particularly compatible with the sedimentational patterns. The Green River can be seen to flow directly across the Uinta basin, where very thick Tertiary fluvial and lacustrine deposits be-tray the Early Tertiary history. Similar relationships exist where the San Juan River obliquely crosses the Early Tertiary fluvial deposits of the San Juan basin. Another problem is that in many cases the present courses of the rivers follow meandering stream patterns reminiscent of the classic meanders of very slug-gish streams flowing across extremely flat land surfaces, such as the lower reaches of the Mississippi River, for example. The question arises as to whether such courses could be maintained in the face of rising landscapes, variable rates of erosion, and variable degrees of equilibrium of the stream gradients. These and other questions plagued the minds of geologists who arrived later.

Erosion Sculpts the Land

The new visitors found these ideas to be untenable, and worried through an alternate hypothesis which they felt answered more of the observable problems than did the "antecedent theory." At the turn of the twentieth century, such eminent geologists as Emmons and Davis believed that the rivers had somehow been let down onto the structures. They supposed further that the uplifts and basins were formed prior to the establishment of the river courses. They felt that the Early Tertiary folds had been buried by later sediments to mask the structure and produce fairly uniform plains across which the meandering streams could flow without reflecting the buried structure.

Their "superposition theory" provided that as the rivers removed the unconsolidated overburden, their beds were lowered into the underlying uplifts and basins, and their paths were necessarily incised into the harder folded rocks. Thus, meandering stream courses were superimposed onto the folded surface beneath and continued to down-cut ever deeper into the underlying strata as the regional uplift of the continental interior progressed.

Geologists argued the pros and cons of these two theories for some fifty years, arriving at various conclusions, which usually encompassed the parts of the two ideas that seemed best suited to solve the particular situation. Recent geomorphologists (geologists who study the origin of the landscape) have generally abandoned both concepts because they do not fully utilize all processes that are known to be presently active.

The later approach to the canyon-cutting enigma was presented in a symposium published by the Museum of Northern Arizona on the "Evolution of the Colorado River in Arizona." The hypothesis presented was based on painstaking study of many facets of the local geology. These include the structure and stratigraphy of the older rocks, Tertiary and Pleistocene stream terraces and lake deposits, and a myriad of other pertinent details. They postulated that in Middle to Late Tertiary time the Colorado River did not cross the Kaibab uplift as it does today. Instead it flowed along the present course of Marble Canyon, then southward along the east flank of the Kaibab uplift as it still does. But then, instead of turning westward to cross the uplift, it continued on south along the present course of the Little Colorado River to terminate in a closed basin in northeastern Arizona. The area is now marked by the deposits of the Late Tertiary Bidahochi Formation, which represents the terminal ponding of the mighty Colorado at that time.

Meanwhile, a drainage system was developing on the west flank of the Kaibab uplift that drained the region from the axis of the uplift westward toward the present-day Lake Mead country. Now, it is a well-known fact that streams commonly extend their length by headward erosion into the source area, where the headwaters accumulate. Modern streams may extend their channels in a headward direction by many feet a year, thereby cutting deeper and deeper into the higher regions that comprise the catchment basins for the snow and rain waters that feed the stream. The western network, known as the "Hualapai drainage system," was sending out headward-growing tributaries into the crestal regions of the Kaibab uplift. One of the more progressive of the tributaries worked its way gradually southeastward into the uplifted plateau. It was eventually able to cross the crest of the great uplift and encounter the canyon of the ancestral Colorado River flowing along the eastern flexure.

What ensued is known as "stream piracy," for the flow of water in one stream is diverted into the channel of another. The Colorado River was thus diverted westward across the Hualapai system, cutting off the flow of water to Lake Bidahochi. As the channel of the newly formed river course deepened its canyon, the flow of water southward was reversed and the Bidahochi section became a northward-flowing tributary to the Colorado—or the Little Colorado River of today. All of this took place in Late Tertiary to early Pleistocene time.

While the "stream-piracy theory" appears to be a practical solution to the origin of the Grand Canyon route, what of the canyons that truncate the other major uplifts? Perhaps with the Grand Canyon study as a model the other canyons will be explained along similar lines of reasoning, after additional detailed studies. At first glance it appears that complex interactions of headward-eroding tributaries and diverted streams might very well explain the present river systems; the main flow being stolen from this channel to service another until a through-flowing course that crosses uplifts randomly is established.

There are some obvious reasons why this theory should be questioned as a universal mechanism for canyon-cutting, and probably a lot that are not so obvious. For example, why are the entrenched meanders of the "Goosenecks of the San Juan River" perched on top of the Monument Upwarp, where headward erosion and piracy should have been most active? Headward erosion of stream courses does NOT produce meanders of this type! What, then, is the connection between the entrenched meanders of this and other localities and the stream-piracy hypothesis? It is doubtful that any single mechanism produced all the features seen on the Colorado Plateau, but careful study and considerable worry will have to be expended on the problems before all the answers will be in.

Regardless of the historical development of the present-day courses of the main rivers, the deepest and most spectacular canyons are today seen to dissect the major uplifts. The process of canyon-cutting is the downward cutting of the channel into older and older strata by the rampaging river and the suspended load of sand and gravel carried along by the waters. The effect is one of perpetual sandblasting of the bottom by the particles in an abrasive action that is not particularly forceful, but relentless in its persistence. If this were the only process active, the canyons would be very deep, with vertical walls.

An accompanying widening process is always in action, however, which broadens the canyons essentially as they are being deepened. This involves several processes, which begin with the tearing away of rock fragments from the exposed canyon walls by the freezing and thawing action working along fractures as well as around sand grains. Torrential rains, which are usually the only form of precipitation in this desert region, assist in tearing away small fragments and further aid in washing the loosened material toward the river below. Along with these processes and sometimes instead of these processes, gravity causes the reduction of canyon walls by moving rock toward the river in the form of landslides and slumps. Animals, of course, are a factor, since they commonly kick and roll rocks down the canyon walls, and man can't resist throwing a few stones down any precipice available. If downward cutting is more effective than lateral erosion, a very steep-walled canyon results. If, on the other hand, the deepening process lags behind for some of many possible reasons, the widening processes will dominate and a broad, shallow canyon will result.

In this manner, vast quantities of sedimentary materials are relocated in a few million years' time, removing the younger strata from the crests of the uplifts and exposing their inner realms. The denudation of the positive structural features is a very slow but effective process, the results of which constitute our present-day scenery across much of the Colorado Plateau.

The Work of Ice

The Pleistocene Epoch of the past million years or so is well known throughout the world as a time of extensive glacial activity. The great ice sheets of the polar regions migrated southward to cover much of the North American continent, reaching far south into the Midcontinent of the United States. Most of Canada and the northern tier of states were overpowered by the continental glaciers, which left behind their legacy of morainal garbage and outwash sediments. The continental glaciers did not reach as far south as the Colorado Plateau.

The higher reaches of the province were attacked by alpine glaciers, however, putting the finishing touches on the San Juan Mountain landscapes as well as the higher, laccolithic desert ranges. The San Juan Mountains were largely transformed from a rolling mountainous upland to an alpine wonderland by the glacial processes. Consequently, the southwestern Colorado uplands provide classic examples of the effects of glacial erosion on mountainous regions.

Glaciers form when the average annual snowfall exceeds the rate of average melting so that an accumulation of snow builds up from year to year. To initiate a glacial cycle, the average annual temperature must be lowered sufficiently to reduce the melting rate. This may require a change of only a couple of degrees on the average, and can result from increased elevation or from climatic changes that may be caused by any one of many factors. In the Pleistocene "ice ages" the average temperatures of much of the world were reduced, and the central United States had just been regionally uplifted. So the reasons for the cooling trend are fairly obvious. Snow began to pile up in the more-protected, north-facing slopes and canyons until they became so overdeepened that they started sliding downhill under the forces of gravity. When a snow pack moves under its own weight it is called a "glacier."

Generally speaking, snow packs of this size are altered to ice internally from a more or less continual process of metamorphism that starts when the snow hits the ground. The first step is that the delicate crystalline projections on the snowflakes disappear by melting and reforming around the nucleus of the crystal to form a granular texture. This so-called "firn," or "corn snow," can be seen on any snow field in the spring as the snow begins to compact. Through further and continual compaction and recrystallization of the snow grains, the mass eventually intergrows to the constituency of ice. Ice will flow by plastic deformation of the ice crystals as the glacier begins to crawl down slope under the effects of gravity.

Glacial erosion is far more effective than other normal processes. It begins in the catchment basin for the snow pack, where the snow and ice freezes on and around the rock with which it comes into contact. As the glacier moves off down the slope, it plucks the underlying rocks from the hillside and carries them along in the glacial ice. This very efficiently carves out the headward regions of the catchment area to form rounded depressions known as "cirques." They grow in size as the glacier continues its labors from season to season. The blocks plucked from the cirque floor and walls are carried along in the basal layers of the glacier and act as abrasives to gouge and scour the floor and walls of the valley down which the ice flows. As the valley is thus sandpapered savagely, it is straightened and worn into a U-shaped cross section that flattens the bottom and over steepens the canyon walls. The steep canyon walls then begin to supply additional debris to the glacier in the form of talus and landslide boulders, which are also incorporated into the ice or ride piggyback on the glacier's surface for a free trip down the valley. Some of the new debris pays its way by aiding in the abrasion of the canyon walls and floor as it is dragged along in the ice.

As adjacent canyons and cirques are widened, they narrow the intervening ridges rapidly and soon produce narrow, knife-edge ridges, or "aretes," between the drainages. When two or more cirques go to work on different sides of a mountain peak, they eventually carve its upper regions into a pyramidal shape, making a sharp summit that is called a "matterhorn" or simply "horn," for the magnificent example of the form represented in the Matterhorn of the Swiss Alps. Thus, distinctive topographic features are formed repeatedly by the process of glacial erosion. These are, in summary, hollowed-out cirques at the head of the glacier, U-shaped valleys along which the glacier flows, striated and polished rock surfaces, aretes between the glaciated valleys, and horns between the heads of two or more adjacent glaciers.

Glacial action is not entirely destructive, however, as the debris that the ice transports must be dumped somewhere. As the tongue of the glacial ice moves down slope into warmer and warmer environments, the ice necessarily begins to melt. Eventually the ice melts away at the front as fast as it is being supplied from the mountains. Thus, the terminus of the glacier is established and remains at a fixed point so long as an equilibrium between ice flow and melting is maintained. The load of boulders carried within and on top of the glacier melts out at the snout of the glacier to be piled up in ridges of precariously balanced boulders. The process resembles an endless conveyor belt, with the glacier picking up rocks from its cirque and along its route and dumping them down-valley in a continuous and endless process. The boulder accumulations are called "moraines," and are further distinguished by the location of their formation into: 1. "terminal moraines" if they form at the terminus of the glacier, 2. "lateral moraines" if they form along the margins of the glacier, and 3. "ground moraines" if they accumulate beneath the ice.

The effects of Pleistocene glaciation in the San Juan Mountains is felt mainly in the erosional scars left by the now-melted glaciers. Most of the main valleys that plunge from the high country are the classic U-shaped and highly straightened profiles that form only from glacial modification. The higher tributary valleys are also usually U-shaped in cross section and hang suspended in a perched position where the lesser glaciers were unable to cut as deeply and rapidly as the main trunk flows to which they were eventually fused. These typical hanging valleys generally lead directly up-valley into the cirques that pock the faces of the upland regions, and form beautiful horn peaks from the highest promontories. Virtually every exposed rock surface in this realm is highly polished and striated by the abrasive forces of the glacier and its tools. From a vantage point such as at Molas Pass on U.S. Highway 550 between Durango and Silverton, all these features can be viewed in one sweeping panorama of the skeletal remains of a once-proud glacial system.

The depositional moraines are found only in the lower elevations, where the glaciers were finally conquered. Within the city limits of Durango, at the base of the once glaciated Animas Canyon, one can readily see the terminal and lateral moraines of the alpine glaciers. They form low ridges of gravel and boulders that parallel the canyon walls only to swing suddenly across the valley in a great U-shaped deposit that partially dams the valley floor in the northern districts of town.

Four glacial advances are recorded in these glacial deposits. The ridges form beautiful perches for the construction of homes, providing excellent views of the scenery, but the boulders raise havoc in the planting and development of lawns. It has now become fashionable to level these moraines for the sites of housing developments in Durango. Not only are the moraines destroyed, but the houses sink slowly into the disturbed Pleistocene deposits. The terminal moraines have now been nearly destroyed by suburban developments.

Fiery Contrast

As the glaciers tore away at the frosty heights of the San Juan Mountains and some of the higher laccolithic ranges such as the La Sals and the La Platas, local stretches of the desert were being scorched and cauterized by lava flows. While the volcanism was not nearly so widespread as in the earlier days of the Tertiary, there were a few scattered localities where the molten torrents were fearsome indeed.

One of the best-known places to see the results of very recent volcanic activity is the Sunset Craters National Monument, a few miles northeast of Flagstaff, Arizona. There the cinder cones and lava flows have been so little scarred by erosion that it seems as though the eruptions must have occurred only a couple of weeks ago. The shape of the parent cinder cone is essentially unaltered from the time of the last eruptions, and all the characteristic features and flow structures of lava flows can be viewed in perfectly preserved form. In fact, one gets the feeling that the next eruption may be only a few minutes away. Several satellite cinder cones in the area are being torn asunder for use as road metal.

Very recent eruptions have also plagued the west end of the Grand Canyon. These flows are rather difficult to view except from river boats, but they are none the less spectacular. Near the mouth of Toroweap Canyon a number of lava flows can be seen to flow downward over the cliffs from the parent vents perched high on the canyon walls. One of the larger of these flows reached the bottom of the canyon some 1.2 million years ago as dated by radioactive potassium-argon techniques. It filled the canyon to a height of 550 feet to form a very effective natural dam. In the million years since the lava dam was formed, the Colorado River has carved its way through the lavas and an additional fifty feet of Paleozoic strata.

Although the lava flows themselves are very interesting, this situation also gives some insight into the rates of erosion of the Grand Canyon and dates the cutting of the canyon to some extent. The canyon was within fifty feet of its present depth at the time the lava dam was formed, and consequently the canyon was already virtually as deep as at present at the beginning of Pleistocene time. The deepening of the canyon has progressed at a rate only about an inch every two thousand years since the lavas reached the canyon bottom. This is a very slow rate indeed, and signifies the importance of vast intervals of geologic time in the formation of the features we see today.

Other regions of Pleistocene volcanism are scattered along the southern margin of the Colorado Plateau to the east of the two regions already discussed. Very well preserved volcanic features of just about every shape and description can be seen between Show Low and St. Johns, Arizona, and again south of the Zuni Mountains uplift in the vicinity of El Morro National Monument. The recency of these eruptions is evident by the common occurrence of lava flows filling the bottoms of the present-day valleys and draws as well as the marvelous preservation of the flow features and cinder cones. These flows fill the valleys surrounding most of the Zuni Mountains and reach northward as far as "The Malpais" near Grants, New Mexico.

Somewhat more-prominent volcanos formed toward the northeast of Grants at this time. By repeated volcanic eruptions Mount Taylor attained an elevation of more than eleven thousand feet immediately northeast of Grants to become a high mountain landmark; the Sacred Mountain of the south, according to Navajo legend. Another volcano of probable Late Tertiary or

Aerial view towards the west across the Grand Canyon in the vicinity of Vulcan's Throne (right-center); Mt. Dellenbaugh at upper right on horizon. Outcropping strata are: Redwall Limestone (Mr), Callville Formation (IPc), and Esplanade Sandstone (Pes). Photo by W. K. Hamblin.

early Pleistocene age grew into a delightful wooded upland just north of Jemez Springs and west of Los Alamos, New Mexico. The large volcano, known as Jemez Caldera, spewed ash and lava over a considerable area of the northern New Mexico countryside, and also grew to an elevation in excess of eleven thousand feet in the form of Redondo Peak, on the caldera rim.

The last great geologic time interval before the Recent, then, was not only destructive in the intense erosional modification of the landscape, but was also constructional in the form of volcanic deposits. The nature of the rock layers, both sedimentary and volcanic, their structural configuration, and the effectiveness of erosion were all contributing factors in the formation of the colorful geologic fairyland we know today as the Colorado Plateau.

PART THREE
Enter Man

Major John Wesley Powell, who lost his right arm
in the Civil War, enthusiastically provides information
to a Paiute Indian, Tau-gu, in southern Utah.
Photo probably taken in about 1873;
courtesy U.S. Geological Survey.

CHAPTER ELEVEN

INDIANS AND EXPLORERS

'ANAASÁZÍ—THE ANCIENT PEOPLE

T he Colorado Plateau with its magnificent deep canyons and towering peaks has not been an environment particularly hospitable to man. It is a land of extremes: flat-topped plateaus and mesas or vertical-walled canyons; ice-draped peaks or desert sands; unbearable heat or arctic blizzards; great aridity or thundering rapids. As a consequence, the earliest inhabitants never really flourished here, and modern man is just beginning to "discover" this enchanted land.

Who the first people were and when they arrived in the Plateau country remains a mystery. The oldest known artifacts found in caves in Grand Canyon are split-willow figurines, probably left by visitors from the Great Basin country to the west, that date back to around 2000 B.C. Why people visited the depths of the Canyon for their religious rites is anybody's guess. There is no further evidence of human visitation until about A.D. 1, when Indians, again probably from the west, were known to have been wandering about the high desert country. These people, the Basketmakers, are named for the numerous examples of basketry they left. They apparently lived on wild corn and squash, but they were basically hunters of bighorn sheep, deer, and smaller native wildlife. Sometime around A.D. 500 these wandering people began to settle down in small villages and became primitive agriculturalists, developing new varieties of corn and growing beans. It is a good bet that the Basketmakers were the forerunners of the first urban settlers in the Four Corners area that modern Navajos call the 'Anaasází (Anasazi), the "Ancient Enemies."

A foremost archeologist in the region, Robert Euler, describes the development of the Anasazi culture in simple terms: "The start of village life was made with groupings of circular pit houses. They manufactured fired pottery and supplemented spearthrowing with the use of the bow and arrow to kill game.

"This stable economic base, combining hunting and gathering with the beginnings of diversification of crops, allowed them to intensify and expand their territory and cultural achievements. After learning to grow cotton, they added spinning and loom weaving to their repertory. Now they wore fabric clothing and added objects of personal adornment. They learned to make exquisite pottery. Thus, midway through the eleventh century A.D., Pueblo culture flowered into its classic traditions. This period, continuing into historic times, saw the rise of great communal pueblos, up to five stories high; these high rise apartment dwellings housed several hundred people and community living must have become highly organized. Evidence indicates that authority was centered in a theocratic hierarchy of priests."

The major Anasazi villages were centered at what is now Mesa Verde National Park near Cortez, Colorado. There magnificent "condominium complexes" were constructed beneath the natural protection of overhanging cliffs of the Cliff House Sandstone (Late Cretaceous) in grottos that provide protection from winter snows and summer sun. These sites usually contain springs that emerge from the contact between the Cliff House and the underlying Menefee Formation. The location of most of the ruins suggests that protection from marauding neighbors was of similar importance in site selection. The suburbs of the main village complex extend for tens of miles in all quadrants from Mesa Verde. Interesting as it may be, there are only few and very small ruins in the deep canyons of the San Juan River and Cataract Canyon on the Colorado River. Although water was readily available all year round and protection was superb, the canyons were apparently too inhospitable and farm land was too limited for communal development.

The thriving culture was plagued with a sustained drought around the close of the thirteenth century, and the Anasazi disappeared from the plateaus, mesas, and canyons of the Four Corners region. They headed south and southwest, into the Rio Grande and Little Colorado drainages, and the Hopi country. They were undoubtedly the direct ancestors of the modern Hopi people of the Black Mesa area and of the various Pueblo people of central New Mexico.

Meanwhile, the Cohonina peoples were inhabiting the area around the south rim of Grand Canyon and northward to the lower Cataract Canyon country. They mimicked the Anasazi in many ways, but never reached the cultural zenith set by their Four Corners neighbors. After A.D. 1300, other, more nomadic groups of Indians entered the Plateau country, giving rise to the Havasupai, Walapai, and southern Paiute tribes; apparently most were descendants of the Cerbat people of the western plateaus and canyons.

Dineh - the People

At a time now lost in antiquity, bands of roving hunters and gatherers began wandering into the Colorado Plateau country. Some came in small groups of family size, some by tens and twenties, some even as loners, in search of food and the "good life" of the next valley. They came by as many routes as there were paths and deer to follow. Most came either from the Athapascan-speaking region of British Columbia or as second or third generation descendants of those peoples. Some began more or less to settle down in the San Juan River Valley around present-day Farmington and Aztec, New Mexico, in what might be called Old Navajoland. These people, relatively nomadic by nature, were in no way akin to the Anasazi, although they were undoubtedly present prior to the great Anasazi exodus. Other groups remained almost totally nomadic, keeping to the mountains where game was plentiful; later they would be distinguished from their Navajo cousins as "Apaches."

There is some evidence to suggest that the closely related Navajos and Apaches were in the vicinity by about A.D. 1100, but it has not been determined with certainty that Navajos were living in their traditional hogans until the middle of the sixteenth century. By that time the Navajos were growing corn. "They felt that their real life had begun when they became corn growers," writes Ruth Underhill. "So it had. If the Navaho had not become corn growers and, later on, sheep raisers, they would have remained just one more little band of the Apache, the Enemy."

The Navajo grew from these humble beginnings to become the largest Indian tribe in North America, both in numbers and in the size of their present-day reservation. The present-day reservation sprawls over an area of northwest New Mexico, northeast Arizona, and southeastern Utah as large as the combined states of Rhode Island, Vermont, Connecticut, and New Hampshire. The lands thus encompassed contain major deposits of oil, gas, coal, and uranium, replete with the management problems commonly associated with mineral wealth.

The Navajos have been accustomed to surviving from what the land provided and have lived in harmony with nature. As a result, the problems of managing the natural resources they inherited with the reservation boundaries have perplexed tribal leaders and tribal councils for generations. Combined with the difficulties of coping with the white man's world, the ever-increasing demand for their natural wealth has become a major concern of the Dineh (the Navajos' word for the Navajo people).

Living in harmony with nature is the basis for Navajo religion; living in harmony with fellow human beings has not been an attribute of the Dineh. In many cases the problems have been caused by hard times and outside provocation. There have apparently always been disputes with other neighboring tribes, especially the Utes to the north. Raiding parties were reportedly crossing the San Juan River in both directions well into the nineteenth century. "White eyes" got into the skirmish in 1846, when the "Blue Coats" wrested New Mexico from the Mexicans, and tried to persuade the Navajo clans to stop raiding Mexican farmers along the Rio Grande Valley and sign a peace treaty. A few agreed, but others, like Manuelito, wanted nothing but the status quo: War! After all, if there was hatred between the Mexicans and the Navajos, what business was it of the "Blue Coats"? The raids continued. Between 1846 and 1850 Dineh helped themselves to over "12,000 mules, 7,000 horses, 31,000 cattle and 450,000 sheep. These were from the Rio Grande villages where a United States marshal could keep count."

In the ensuing skirmishes, the "Blue Coats" shot and killed a Navajo head man, Narbona, and the Navajos became the enemies of the whites for life. And the fighting continued, especially for Manuelito, the son-in-law of Narbona, and his followers. In an effort to bring the "war" to a halt, the whites established a military outpost in the heart of the Navajo country and called it Fort Defiance. Although the whites and their Hopi, Zuni, and Ute scouts and aides had muskets, the Navajos had only bows and arrows and spears to continue the wars. Nevertheless they seemed to have won when Fort Defiance was abandoned in 1861, at the beginning of the Civil War.

Raiding of villages, farms, and the trails through the Southwest continued during the Civil War. The Union Army had to take severe measures to halt the annoyances, once and for all. Kit Carson, a well-known and respected guide for explorers and the military, and a group of untrained, largely Mexican, volunteers were ordered to round up the Navajo and Apache marauders and forcibly place them in concentration camps, known as reservations.

By March 1863, 400 Mescalero Apaches had been sent from their raiding grounds forever to live in inhospitable and unfamiliar surroundings. Although most of the Indian reservations were set aside in Indian Territory in Oklahoma, it was decided that the Navajos and Apaches were to be taken to Fort Sumner, known to the Mexicans as Bosque Redondo, near the Pecos River in eastern New Mexico. The Navajos were asked to move peacefully to the new "promised land," and some who had already turned to farming and were friendly with the whites agreed. Most, however, decided to remain in their homelands and fight.

They fled "like birds into the canyons and among the rocks." Carson and his small army of 700 volunteers had difficulty finding them, even with the help of Mexicans, Utes, and Pueblos, who were only too happy to help round up the Navajos in revenge. The Indians scattered to the four corners of Indian country. Rather than waste years searching out Dineh, Carson decided on a different approach—to destroy all the Navajo cornfields and sheep he could find; their wealth and food supplies. A few small bands were forced to surrender at the reopened Fort Defiance out of hunger, but many still hid, living on roots and seeds, rats and gophers, and nearly freezing through the cold winter. Most eventually took refuge in Canyon de Chelly, where soldiers had not been. Then in January of 1864, Carson decided to enter the canyon. Soldiers destroyed the cornfields and cut down the peach trees in the canyon, forcing the hiding Navajos to surrender because of the bitter cold and hunger. Gradually, the Indians gathered at Fort Defiance, where they were fed and treated surprisingly well.

Then on March 6, 1864, the "Long Walk" of 300 miles to Fort Sumner began; 2,400 Navajo, what sheep and horses they had salvaged, and their Mexican escorts made their way painfully to the reservation. There they were kept in exile for four long years, some 9,000 people living on 40 square miles of land, 6,000 acres of which could be farmed. They were forced to live in hand-dug pithouses, grow crops in a technique hitherto unknown to them, straight rows, and walk miles for firewood. When they tried to slip away, they were brought back. The untamed Comanches raided their herds; the crops failed year after year. After three years of total despair on the part of the Navajos and $10 million of disappointing expenditure by the government, a treaty was signed on June 1, 1868. This established the so-called treaty reservation and the return of the Navajo to at least a part of their native habitat. The new way of life had begun.

(For readers interested in detailed accounts of the early history of the Navajo Nation, the book "Here Come the Navaho" by Ruth Underhill is strongly recommended.)

The Treaty Reservation as established in 1868 included but a fraction of the traditional Navajo Country and excluded Dinehtah: "Old Navajoland." Various modifications to the boundaries by presidential executive orders and congressional acts began in 1878 and continued to as late as 1958, adding many of the traditional lands to the reservation, but still not "Old

Navajoland." Long-standing disputes over the Hopi lands of Black Mesa are now being settled (not to everyone's satisfaction) by the courts (see The Second Long Walk, by Jerry Kammer, 1980).

The transition from the old ways to modern times has been fraught with dispute, wars, and problems with the white man's justice, setting the stage for the modern "energy crunch."

The "White Eyes"

Meanwhile, intruders from another world were encroaching on the Native Americans' land. Within 50 years after Columbus landed in the New World, the Spanish explorers were dazzled by the magnificent view into the Grand Canyon from the South Rim, wondering how to cross the abyss. Although his was not the first Spanish intrusion into the Southwest, Garcia Lopez de Cardenas was ordered to locate the legendary canyon by Coronado, and in late September 1540 the expedition led by Indian guides "discovered" the Canyon. The Spaniards would not visit Grand Canyon again for nearly 250 years.

These European explorers in search of the fabulous Seven Cities of Cibola traveled through parts of the Navajo country between 1535 and 1605, but they did not find the towns with streets paved in gold that they expected. Then in the fall of 1775, Fray Francisco Tomas Garces set out to find a route from New Mexico to California, visiting the Grand Canyon and traveling to Oraibi in 1776, finding the Hopi to be unfriendly, perhaps as a result of a Navajo raid.

Certainly the most famous of the Spanish explorers were Francisco Atanasio Dominguez and Silvestre Velez de Escalante, who explored a large part of the Colorado Plateau in two expeditions of 1765–66 and 1776–77. They followed a most circuitous route in search of a way to California from Santa Fe, New Mexico. The Franciscan Friars passed through western Colorado to the Green River near Vernal, Utah, then westward to about Provo and followed the western margin of the Colorado Plateau southwestward to about St. George. They then turned back toward the east across the Arizona "strip" country, crossing the Colorado River at what is now known as the Crossing of the Fathers (now under Lake Powell), and went back to Santa Fe via Oraibi. They were apparently the first tourists on the Colorado Plateau.

The Dominguez-Escalante expeditions served to open up the vast plateau country and left a heritage of Spanish place names to the region. Names such as San Juan, La Plata, Abajo, La Sal, Las Animas, Colorado, and others apparently were derived from their wanderings. Then in the years following

1829, a trading route was established between Santa Fe and southern California that skirted the plateau country. It was a variation of the explorers' route, now somewhat erroneously called the Spanish Trail, in order to circumvent the hazards of the marauding Navajo and Apache Indians of northern Arizona and New Mexico. That route passed near Cortez, Colorado, crossed the Colorado River at Moab, Utah, the Green River at Green River, Utah, and went on to Salt Lake country. Still the inner canyon realms of the Colorado Plateau were avoided for easier traveling conditions.

During the years 1821–48, when the plateau country was still part of Mexico, American fur men were everywhere they were likely to find good pelts. They ranged from the upper Green River to the lower Gila but found the best hunting in northern Utah and on the western slope of Colorado. They traveled country known only by the Indians and made extensive excursions into the magnificent canyons of the Green, Colorado, and San Juan rivers. They were not doing such wondrous things as "exploring"; they were simply looking for the then lucrative beaver pelts. As a result, little is known of their findings, for they wrote precious little about their travels. One such "explorer," Denis Julien, a French Canadian, probably made the first passage

Inscription left by Denis Julien near the mouth of Hell Roaring Canyon where it enters the Green River upstream from the confluence with the Colorado River.

Indians and Explorers   215

Geologists of the 1921 Triamble expedition plane-table mapping beneath Mexican Hat rock on the San Juan River, southeastern Utah. Photo by H. D. Miser, courtesy of the U.S. Geological Survey.

through Cataract Canyon by boat. He left his trail in the form of inscriptions on the canyon walls in 1836. These hardy men were the first to become familiar with the intimacies of the geography, Indians, and wild life, and the trails they used have become the highways of modern times.

The Scientists

In the early 1860s the Army Topographical Engineers sought railroad routes across the plateau country to California. Edward Beale, John Gunnison, John Charles Fremont, and Amiel Whipple crossed the plateau by routes that were later to become lines of the Denver and Rio Grande Western and the Santa Fe railroads. The most notable exploration was made in 1859 by Capt. J. N. Macomb, who traveled from Santa Fe through southeastern Utah's spectacular Canyonlands region to reach a point near the confluence of the Green and Colorado rivers.

Exploration for its own sake continued with the two expeditions (1869 and 1871–72) of Maj. John Wesley Powell down the full length of the Green and Colorado rivers, using little wooden boats and motley crews. The major took a deep interest in the Indians of the region, as well as the geology, and later founded the Bureau of American Ethnology and was the second director of the U.S. Geological Survey. Powell's survey and others headed by George M. Wheeler and F. V. Hayden mapped and studied almost all of the Colorado Plateau. The survey of Clarence King passed through northern Utah's Uinta Mountains and went on to the west into Nevada.

The reports, atlases, and scientific papers published by these surveys constitute a primary body of geological, geographical, ethnological, and biological literature of the region that runs to dozens of titles. These publications

Glen Canyon dam near the Arizona-Utah border forms Lake Powell, named for John Wesley Powell, first Colorado River explorer. The rocks are massive sandstone beds of Jurassic age.

did much to attract public attention to the plateau, and many of them are of value, even today. Several hundred photos were taken in conjunction with these surveys, and today the exquisite photos painstakingly made with wet glass plates are being studied and rephotographed in detail.

The Colorado Plateau Today

Still, the region was little known to modern man until within the past few decades, when economic stimuli lured the modern-day explorers off the highways and into the desert. Even with modern tools and techniques and economic incentive, the Colorado Plateau has remained essentially untamed. The extremely rugged terrain has been the primary deterrent, for the giant hogbacks, or "reefs," that guard the major uplifts, and the seemingly bottomless canyons that dissect the strata, form virtually impassable barriers to modern forms of travel. The way that John Wesley Powell traveled a hundred years ago in his little wooden river boats is still almost the only way the canyons can be explored. The rugged terrain can be seen better today from the vantage point of a low-flying airplane, but it is not easy to collect fossils and arrowheads from the air, and it would be a very long walk if the motor were to quit. The modern bulldozer and jeep have been the most effective tools in opening up this wilderness, but by far the greatest share of the province remained untrod by white men until it became economically justifiable to carve trails from the vertical cliffs in the search for oil and uranium.

Today, jeep trails and tiny, rocky landing strips provide rough access to much of the less rugged country. Still, a four-wheel-drive vehicle, a constitution of steel, and a thorough knowledge of desert survival are required to experience the majesty of the Colorado Plateau; only by helicopter or river boat can the innermost secrets be known. So this is still one of the great wilderness regions that have only barely been scratched by the progress of man.

Most of the province cannot even be imagined from the modern network of highways. Paved and otherwise improved roads only skirt the vast canyon regions, as do the railroads, for even with present-day skills the canyons cannot be tamed. The modern paths of travel tiptoe along the margins of the rough country, but do not enter. No highways or railroads cross the Canyonlands or Grand Canyon regions, so that only the more hardy and adventuresome traveler can visit these spectacles. "You can't get there from here," is a commonly heard phrase in this country.

Uranium fever in the early 1950s was responsible for the construction of

Lower Cataract Canyon as it appears after inundation by Lake Powell in Glen Canyon Recreation Area, southern Utah. Canyon walls are of Pennsylvanian and Permian age rocks.

most of the jeep trails into the rough country. Today, the outcrops of the Jurassic and Triassic rocks and the best access routes to these strata are well marked by the tortuous little trails. Most of the tiny airstrips were scratched from the desert to supply the best prospecting regions, but these have long been abandoned and left to decompose naturally. Many fortunes were made and lost at the ends of the rocky trails and in the stock market.

Some highly prosperous mining regions emerged from the chaos. The White Canyon and Monument Valley districts on the southern Monument Upwarp produced economic profits from the Shinarump and Moss Back members of the Triassic Chinle Formation. The yellow riches come from the Salt Wash

Member of the Jurassic Morrison Formation in the Uravan Mineral Belt, where the Paradox, Gypsum, and Lisbon valleys, and other nearby areas, were among the richest uranium finds in the world. One can usually identify the Shinarump and Salt Wash stratigraphic units by the intensity of roads and bulldozed prospect pits along their outcrops. More recently the Grants, New Mexico, uranium districts were developed in upper members of the Morrison Formation, and are the country's biggest potential source of uranium. All was abandoned when the demand for uranium went bust. As future needs for uranium and vanadium arise, the Colorado Plateau will again come to life.

Some of the same regions also host the lion's share of the petroleum reserves on the Colorado Plateau. The Lisbon Valley structure, for example, is not only the host to the Big Indian uranium district, which is among the very richest, but also contains the prolific Lisbon Valley oil field, which produces from the Mississippian Redwall Limestone deep within the substructure of the salt-intruded anticline. Other Mississippian oil production in the vicinity came from the McIntyre Canyon field a short distance to the southeast, the Big Flat field near Dead Horse Point west of Moab, Utah, and another small discovery at the Salt Wash field a few miles farther northwest.

The bulk of the petroleum wealth of the Colorado Plateau is presently being exploited in the Four Corners region. There, millions of barrels of the black gold are being produced from algal bioherms of Middle Pennsylvanian age in southeasternmost Utah in the Aneth Field and numerous other, smaller, outlying fields. Discovery of commercial oil fields is still in progress in that region, although exploration activity has slowed to a snail's pace since the mid-1980s.

The oil exploration world was stunned by the discovery of commercial quantities of oil in Tertiary igneous sills that have intruded the Pennsylvanian strata in the north end of the Chuska Mountains just south of Four Corners. Igneous rocks are the last place an oil geologist looks for oil reservoirs. The rocks are usually nonporous and impermeable, and the heat of the intrusion should ordinarily destroy any oil in the adjacent rocks. Yet the oil is flowing.

Some oil but primarily gas is being produced from the sandstone formations of Cretaceous age in the San Juan basin of northwestern New Mexico. These occurrences are also unusual in that the petroleum is coming from downfolded strata only a short distance from their outcrops.

Still another area of petroleum wealth is located in the Uinta basin of northeastern Utah, near Vernal, where very heavy, black oil is being pumped from the Tertiary lake beds. Gilsonite, a black, brittle mineral formed by the loss of the volatile fractions of petroleum, also occurs in this area in large dikes that penetrated the sedimentary rocks and extend for miles across the countryside exposed at the very surface. This solid petroleum is mined, crushed, and injected into oil pipelines and sent along to the refinery. The largest oil field in Colorado, the Rangely Field, has produced prodigious amounts of oil, mainly from the Pennsylvanian Weber Sandstone in the eastern uinta basin. And amid much controversy, oil shale was developed experimentally and abandoned in northwestern Colorado.

Deposits of lead, zinc, copper, silver, and gold ores lie as dormant wealth in the San Juan Mountains of southwestern Colorado, awaiting stable high metal prices that will make them again economically viable. The bulk of the ore deposits occur in the Silverton to Ouray region, where their discovery in the late 1800s and early 1900s opened up the mountainous country with innumerable horse (and now jeep) trails. The ancient access routes to the original mines and prospects are now utilized by hundreds of jeep enthusiasts annually. The early silver fever, therefore, helped open up the West, not only to the early pioneers but also to the modern explorer.

Scenery is still the biggest and best-selling commodity on the plateau, in spite of the natural petroleum and mineral riches. And the geologic history is in large part responsible for the magnificent country. The numerous national parks and monuments, which depend on the interrelationships of scenery and geology, attest the visual and geologic beauty of the region. A new expansive region, the Grand Staircase-Escalante National Monument was established in 1999. Yet some of the very best in scenic and recreational attractions lie outside park boundaries. The political climate of the new millennium, however, promises to create more park protection. The dark, laminated canyons with their tortuous tributaries and unchallenged spires and buttes provide an air of mystery that is inescapable and unconquerable to the sensitive tourist. The rugged desert beauty and the unbelievable exposures of the strata make this a geologist's paradise.

The region is now largely uninhabited except by Indians, tour guides, and tourists. It is so unexplored by the public that it has not yet been buried under the "beer-can conglomerate" so common in our modern world. The economic, recreational, and cultural potential of the region is too little developed to permit predictions—no end to the excitement and enchantment of the Colorado Plateau is in sight.

.

GLOSSARY

ANGULAR UNCONFORMITY
An unconformity or break between two series of rock layers such that rocks of the lower series underlie rocks of the upper series at an angle; the two series are not parallel. The lower series was deposited, then tilted and eroded prior to deposition of the upper layers.

ANTICLINE
An elongate fold in the rocks in which sides slope downward and away from the crest; an upfold.

ARKOSE
A sandstone containing a significant proportion of feldspar grains, usually signifying a source area composed of granite or gneiss.

BASEMENT
In geology, the crust of the Earth beneath sedimentary deposits, usually, but not necessarily, consisting of metamorphic and/or igneous rocks of Precambrian age.

BASEMENT FAULT
A fault that displaces basement rocks and originated prior to deposition of overlying sedimentary rocks. Such faults may or may not extend upward into overlying strata, depending upon their history of rejuvenation.

BASE LEVEL
The level, actual or potential, toward which erosion constantly works to lower the land. Sea level is the general base level, but there may be local, temporary base levels such as lakes.

BENTONITE
A rock composed of clay minerals and derived from the alteration of volcanic tuff or ash.

BRACHIOPOD
A type of shelled marine invertebrate now relatively rare but abundant in earlier periods of Earth history. They are common fossils in rocks of Paleozoic age. Brachiopods have a bivalve shell that is symmetrical right and left of center.

BRYOZOA
Tiny aquatic animals that build large colonial structures that are common as fossils in rocks of Paleozoic age.

CARBON-14 DATING
OR RADIOCARBON DATING
A method of determining an age in years by measuring the concentration of carbon-14 remaining in formerly living matter, based on the assumption that assimilation of carbon-14 ceased abruptly at the time of death and that it thereafter remained a closed system. A half-life of 5570+/-30 years for carbon-14 makes the method useful in determining ages in the range of 500-40,000 years.

CEPHALOPOD
Marine mollusks that secrete shells that are chambered, usually coiled in a planospiral, but occasionally straight, with the main body of the animal housed in the last open chamber; they maintain buoyancy with gas filling the enclosed chambers and swim by jetting fluid; modern examples are the nautilus and squid. Ammonitic cephalopods have complexly crinkled chamber walls that may be used to distinguish species. They are especially useful in dating Mesozoic rocks.

CHERT
A very dense siliceous rock usually found as nodular or concretionary masses, or as distinct beds, associated with limestones. Jasper is red chert containing iron-oxide impurities.

CLASTIC ROCKS
Deposits consisting of fragments of preexisting rocks; conglomerate, sandstone, and shale are examples.

CONGLOMERATE
The consolidated equivalent of gravel. The constituent rock and mineral fragments may be of varied composition and range widely in size. The rock fragments are rounded and smoothed from transportation by water.

CONODONTS
Toothlike microfossils made of amber-colored calcium phosphate that may occur singularly or in mixed assemblages of variously shaped parts. They occur only in fine-grained rocks of marine origin and are believed to be derived from extinct annelid worms. As they evolved very rapidly, they are useful in dating and correlating rocks of Paleozoic and Mesozoic age.

CONTACT
The surface, often irregular, which constitutes the junction of two bodies of rock.

CONTINENTAL DEPOSITS
Deposits laid down on land or in bodies of water not connected with the ocean.

CORRELATION
The process of determining the position or time of occurrence of one geologic phenomenon in relation to others. Usually it means determining the equivalence of geologic formations in separated areas through a comparison and study of fossils or rock peculiarities.

CRINOID
Marine invertebrate animals, abundant as fossils in rocks of Paleozoic age. Most lived attached to the bottom by a jointed stalk, the "head" resembling a lily-like plant, hence the common name "sea lily."

DIATREME
A breccia-filled volcanic pipe that was formed by a gaseous explosion.

DIKE
A sheetlike body of igneous rock that filled a fissure in older rock while in a molten state. Dikes that intrude layered rocks cut the beds at an angle.

DISCONFORMITY
A break in the orderly sequence of stratified rocks above and below which the beds are parallel. The break is usually indicated by erosional channels, indicating a lapse of time or absence of part of the rock sequence.

DOLOMITE
A mineral composed of calcium and magnesium carbonate, or a rock composed chiefly of the mineral dolomite, formed by alteration of limestone.

DOME
An upfold in which strata dip downward in all directions from a central area; the opposite of a basin.

EOLIAN
Pertaining to wind. Designates rocks or soils whose constituents have been transported and deposited by wind. Windblown sand and dust (loess) deposits are termed eolian.

EROSIONAL UNCONFORMITY
A break in the continuity of deposition of a series of rocks caused by an episode of erosion.

EXTRUSIVE ROCK
A rock that has solidified from molten material poured or thrown out onto the Earth's surface by volcanic activity.

FACIES
Generally, this term refers to a physical aspect or characteristic of a sedimentary rock, as related to adjacent strata. It is usually applied to distinguish different aspects of the sediments in time-equivalent or laterally continuous beds. For example, the white sandstone facies of the Cedar Mesa Sandstone changes laterally to the age-equivalent red arkosic sandstone facies of the Cutler Group in Canyonlands Country. Such a change from one aspect to another is called a facies change.

FAULT
A break or fracture in rocks, along which there has been movement, one side relative to the other. Displacement along a fault may be vertical (normal or reverse fault) or lateral (strike-slip or "wrench" fault).

FORAMINIFERA
Generally microscopic one-celled animals (Protozoa), almost entirely of marine origin, with sufficiently durable shells capable of being preserved as fossils. They are usually abundant in marine sediments, and are sufficiently small to be retrievable in drill cuttings and cores.

FORMATION
The fundamental unit in the local classification of layered rocks, consisting of a bed or beds of similar or closely related rock types, and differing from strata above and below. A formation must be readily distinguishable, thick enough to be mappable, and of broad regional extent. A formation may be subdivided into two or more MEMBERS, and/or combined with other closely related formations to form a GROUP.

FUSULINDS

Small spindle-shaped Foraminifera occurring as elongate chambers enrolled into complex internal forms that resemble a jellyroll. They are found only in marine rocks of Pennsylvanian and Permian age where they are excellent fossils for dating and correlating sedimentary rocks because of their rapid evolutionary history.

GEOLOGIC MAP

A map showing the geographic distribution of geologic formations and other geologic features, such as folds, faults, and mineral deposits, by means of color or other appropriate symbols.

GNEISS

A banded metamorphic rock with alternating layers of usually elongated tabular, unlike minerals.

GRANITE

An intrusive igneous rock with visibly granular, interlocking, crystalline quartz, feldspar, and perhaps other minerals.

IGNEOUS ROCK

Rocks formed by solidification of molten material (magma), including rocks crystallized from cooling magma at depth (intrusive), and those poured out onto the surface as lavas (extrusive).

INTRUSIVE ROCK

Rock that has solidified from molten material within the Earth's crust and has not reached the surface; it usually has a visibly crystalline texture.

LIMESTONE

A bedded sedimentary deposit consisting chiefly of calcium carbonate, usually formed from the calcified hard parts of organisms.

MASSIF

A massive topographic and structural uplift, commonly formed of rocks more rigid than those of its surroundings. These rocks are commonly protruding bodies of basement rocks, consolidated during earlier orogenies.

METAMORPHIC ROCK

Rocks formed by the alteration of preexisting igneous or sedimentary rocks, usually by intense heat and/or pressure, or mineralizing fluids.

MINETTE

A dark-colored intrusive igneous rock primarily composed of biotite phenocrysts (enlarged crystals) in a groundmass of orthoclase feldspar and biotite; it is commonly found in dikes associated with diatremes.

MORAINE

A mound, ridge, or other distinct accumulation of unsorted, unstratified drift, predominantly a heterogeneous mixture of mud, sand, gravel, and boulders, deposited by the melting of glacial ice.

OROGENY

Literally the process of formation of mountains, but practically the processes by which structures in mountainous regions were formed, including folding, thrusting, and faulting in the outer layers of the crust, and plastic folding, metamorphism and plutonism (emplacement of magma) in the inner layers. An episode of structural deformation may be called an orogeny, e.g., the Laramide orogeny.

SANDSTONE

A consolidated rock composed of sand grains cemented together; usually composed predominantly of quartz, it may contain other sand-size fragments of rocks and/or minerals.

SCHIST
A crystalline metamorphic rock with closely spaced foliation (platy texture) that splits into thin flakes or slabs.

SEDIMENTARY ROCK
Rocks composed of sediments, usually aggregated through processes of water, wind, glacial ice, or organisms, derived from preexisting rocks. In the case of limestones, constituent particles are usually derived from organic processes.

SHALE
Solidified mud, clays, and silts, that are fissile (split like paper) and break along original bedding planes.

SILL
A tabular body of igneous rock that was injected in the molten state concordantly between layers of preexisting rocks.

STRATIGRAPHY
The definition and interpretation of layered rocks, the conditions of their formation, their character, arrangements, sequence, age, distribution, and correlation, using fossils and other means.

STRATUM
A single layer of sedimentary rock, separated from adjacent strata by surfaces of erosion, non-deposition, or abrupt changes in character. Plural: strata.

SYNCLINE
An elongated, troughlike downfold in which the sides dip downward and inward toward the axis.

TECTONIC
Pertaining to rock structures formed by Earth movements, especially those that are widespread.

TRILOBITE
A general term for a group of extinct animals (arthropods) that occurs as fossils in rocks of Paleozoic age. The fossils consist of flattened, segmented shells with a distinct thoraxial lobe and paired appendages, usually found as partial fragments.

TYPE LOCALITY
The place from which the name of a geologic formation is taken and where the unique characteristics of the formation may be examined.

UNCONFORMITY
A surface of erosion or non-deposition separating sequences of layered. rocks.

BIBLIOGRAPHY

Abbot, W. O.; Liscomb, R. L. "Stratigraphy of the Book Cliffs in East Central Utah." Intermountain Association of Petroleum Geologists, *7th Field Conference Guidebook*, pp. 120–23, 1956

Akers, J. P.; Cooley, M. E.; Repenning, C. A. "Moenkopi and Chinle Formations of Black Mesa and Adjacent Areas," New Mexico Geological Society, *9th Field Conference Guidebook*, pp. 88–94, 1958

Allen, J. E.; Balk, R. "Mineral Resources of Fort Defiance and Tohatchi Quadrangles, Arizona and New Mexico." New Mexico Bureau of Mines and Mineral Resources, *Bulletin 36*, 1954

Armstrong, A. K. "Meramecian (Mississippian) Endothyrid Fauna from the Arroyo Peñasco Formation, Northern and Central New Mexico." *Journal of Paleontology*, v. 32, No. 5, pp. 970–76, 1958

___ "Biostratigraphy and Carbonate Facies of the Mississippian Arroyo Peñasco Formation, North-Central New Mexico." *New Mexico Bureau of Mines Resources Memoir* 20, 1967

Armstrong, A. K.; Holcomb, L. D. "Stratigraphy, Facies and Paleotectonic History of Mississippian Rocks in the San Juan Basin of Northwestern New Mexico and Adjacent Areas." New Mexico Geological Society, *40th Field Conference Guidebook*, pp. 159–166, 1989

Armstrong, A. K.; Mamet, B. L. "Biostratigraphy and Paleogeography of the Mississippian System in Northern New Mexico and Adjacent San Juan Mountains of Southwestern Colorado." In J. E. Fassett and H. L. James, eds., New Mexico Geological Society, *28th Field Conference Guidebook*, pp. 111–127, 1977

Atwood, W. W.; Mather, K. F. "Physiography and Quaternary Geology of the San Juan Mountains, Colorado." U.S. Geological Survey Professional Paper 166, 1932

Baars, D. L. "Cambrian Stratigraphy of the Paradox Basin Region." Intermountain Association of Petroleum Geologists, *9th Field Conference Guidebook*, pp. 93–101, 1958

___ "Permian Blanket Sandstones of Colorado Plateau." Symposium of American Association of Petroleum Geologists: *Geometry of Sandstone Bodies*, 1961

___ "Permian Strata of Central New Mexico." New Mexico Geological Society, *12th Field Conference Guidebook*, pp. 113–20, 1962

___ "Permian System of Colorado Plateau." *American Association of Petroleum Geologists Bulletin*, v. 46, pp. 149–218, 1962

___ "Petrology of Carbonate Rocks." *Four Corners Geological Society Symposium: Shelf Carbonates of the Paradox Basin*, pp. 101–29, 1963

___ "Modern Carbonate Sediments as a Guide to Ancient Limestones." *World Oil Magazine*, v. 158, No. 5, pp. 95–100, 1964

___ "Pre-Pennsylvanian Paleotectonics—Key to Basin Evolution and Petroleum Occurrences in Paradox Basin, Utah and Colorado." *American Association of Petroleum Geologists Bulletin*, v. 50, pp. 2082–2111, 1966

___ "Permianland: The Rocks of Monument Valley." In H. L. James, ed., Guidebook of Monument Valley and Vicinity, Arizona and Utah. New Mexico Geological Society, *24th Field Conference Guidebook*, pp. 68–71, 1973

___ "The Colorado Plateau Aulacogen—Key to Continental Scale Basement Rifting. " Podwysocki and Earle, *eds. Proceedings of the Second International Conference on Basement Tectonics*, pp. 157–64, 1976.

___ "Triassic and Older Stratigraphy; Southern Rocky Mountains and Colorado Plateau." In L. L. Sloss, Ed., *Sedimentary Cover, North American Craton, U.S.* Geological Society of America, The Geology of North America, v. D-2, pp. 53–64, 1988

___ "Redefinition of the Pennsylvanian and Permian Boundary in Kansas, Midcontinent, USA." Program and Abstracts, *International Congress on the Permian System of the World*, Perm', USSR, A-3, 1991

___ *The American Alps—The San Juan Mountains of Southwest Colorado*. University of New Mexico Press, Albuquerque, p. 194, 1992

___ *Canyonlands Country—Geology of Canyonlands and Arches National Parks*. University of Utah Press, Salt Lake City, p. 138, 1993

___ "Proposed Repositioning of the Pennsylvanian-Permian Boundary in Kansas." Kansas Geological Survey, *Bulletin 231*, pp. 5–11, 1994

Baars, D. L.; Knight, R. L. "Pre-Pennsylvanian Stratigraphy of the San Juan Mountains and Four Corners Area." New Mexico Geological Society, *8th Field Conference Guidebook*, pp. 108–30, 1957

Baars, D. L.; Parker, J. W.; Chronic, J. "Revised Stratigraphic Nomenclature of Pennsylvanian System, Paradox Basin." *American Association of Petroleum Geologists Bulletin*, v. 51, pp. 393–403, 1967

Baars, D. L.; Seager, W. R. "Depositional Environment of White Rim Sandstone (Permian), Canyonlands National Park, Utah" (abstract). *American Association of Petroleum Geologists Bulletin*, v. 51, p. 453, 1967

Baars, D. L.; See, P. D. "Pre-Pennsylvanian Stratigraphy and Paleotectonics of the San Juan Mountains, Southwestern Colorado." *Geological Society of America Bulletin*, v. 79, pp. 333–50, 1968

Baars, D. L.; Stevenson, G. M. "Permian Rocks of the San Juan Basin." In J. E. Fassett, ed., New Mexico Geological Society, *28th Field Conference Guidebook*, pp. 133–138, 1977

___ "Tectonic Evolution of the Paradox Basin." In D. L. Weigand, ed., *Rocky Mountain Association of Geologists Guidebook*, pp. 23–31, 1981

___ "Subtle Stratigraphic Traps in Paleozoic Rocks of the Paradox Basin." In M. Halbouty, ed., American Association of Petroleum Geologists Memoir 32, pp. 131–158, 1982

Baars, D. L. and 15 others, "Basins of the Rocky Mountain Region." In L. L. Sloss, ed., *Sedimentary Cover, North American Craton, U.S.* Geological Society of America, The Geology of North America, v. D-2, pp. 109–220, 1988

Baars, D. L.; Thomas, W. A.; Drahovzal, J. A.; Gerhard, L. C. "Preliminary Investigations of Basement Tectonic Fabric of the Conterminous USA." in R. W. Ojakangas and others, eds., *Basement Tectonics 10*, pp. 149–158, 1995

Baker, A. A. "Geology and Oil Possibilities of the Moab District, Grand and San Juan Counties, Utah." *U.S. Geological Survey Bulletin 841*, 1933

___ "Geologic Structure of Southeastern Utah." *American Association of Petroleum Geologists Bulletin*, v. 19, pp. 1472–1507, 1935

___ "Geology of the Monument Valley—Navajo Mountain Region, San Juan County, Utah." *U.S. Geological Survey Bulletin 865*, p. 106, 1936

___ "Geology of the Green River Desert—Cataract Canyon Region, Emery, Wayne, and Garfield Counties, Utah." *U.S. Geological Survey Bulletin 951*, p. 122, 1946

Baker, A. A.; Dane, C. H.; McKnight, E. T. "Geologic Structure of Parts of Grand and San Juan Counties, Utah." U.S. Geological Survey Preliminary Map, 1931

Baker, A. A.; Dane, C. H.; Reeside, J. B. Jr. "Paradox Formation of Eastern Utah and Western Colorado." *American Association of Petroleum Geologists Bulletin*, v. 17., pp. 963–80, 1933

___ "Revised Correlation of Jurassic Formations of Parts of Utah, Arizona, New Mexico, and Colorado." *American Association of Petroleum Geologists Bulletin*, V. 31, 1947

Baker, A. A.; Dobbin, C. E.; McKnight, E. T.; Reeside, J. B. "Notes on the Stratigraphy of the Moab Region, Utah." *American Association of Petroleum Geologists Bulletin*, v. 11, pp. 785–808, 1927

Baker, A. A.; Reeside, J. B. Jr. "Correlation of the Permian of Southern Utah, Northern Arizona, Northwestern New Mexico, and Southwestern Colorado." *American Association of Petroleum Geologists Bulletin*, v. 13, pp. 1413–48, 1929

Barnes, H. "Age and Stratigraphic Relations of Ignacio Quartzite in Southwestern Colorado." *American Association of Petroleum Geologists Bulletin*, v. 38, pp. 1780–91, 1954

Bass, N. W. "Correlation of Basal Permian and Older Rocks, Southwest Colorado, Northwest New Mexico, Northeast Arizona, and Southeast Utah." U.S. Geological Survey Oil and Gas Investigations, Preliminary Chart No. 7, 1944

___ "Paleozoic Stratigraphy as Revealed by Deep Wells in Parts of Southwestern Colorado, Northwestern New Mexico, Northeastern Arizona, and Southeastern Utah." U.S. Geological Survey Oil and Gas Investigations, Preliminary Chart No. 7, 1944

Beaumont, E. C.; Read, C. B. "Geologic History of the San Juan Basin Area, New Mexico and Colorado." *New Mexico Geological Society Guidebook, First Field Conference*, pp. 49–52, 1950

Bebenroth, D. L.; Strahler, E. N. "Geomorphology and Structure of the East Kaibab Monocline." *Geological Society of America Bulletin*, v. 56, pp. 107–50, 1945

Berry, E. W. "Cycads in the Shinarump Conglomerate of Southern Utah." *Washington Academy of Sciences Journal*, v. 17, pp. 303–7, 1927

Beus, S. S.; Morales, M. *Grand Canyon Geology*. Oxford University Press, New York and Museum of Northern Arizona, Flagstaff, 1990

Blagbrough, J. W. "Cenozoic Geology of the Chuska Mountains." In F. D. Trauger, ed., New Mexico Geological Society *18th Field Conference Guidebook*, pp. 70–77, 1967

Blakey, R. C. "Stratigraphy and Geologic History of Pennsylvanian and Permian Rocks, Mogollon Rim, Central Arizona and Vicinity." *Geological Society of America Bulletin*, v. 102, pp. 1189-1217, 1990

___ "Supai Group and Hermit Formation." In S. S. Beus and M. Morales, eds., *Grand Canyon Geology*. Oxford University Press and Museum of Northern Arizona Press, pp. 147–182, 1993

Bradish, B. B. "Geology of the Monument Upwarp." Four Corners Geological Society Symposium, pp. 47–60, 1952

Broan, B. "The Cretaceous Ojo Alamo Beds of New Mexico, with Description of the New Dinosaur Genus *Kritossaurus*." *American Museum of Natural History Bulletin*, v. 28, Art. 24, pp. 267–74, 1910

Burbank, W. S.; et al. "Ouray District, Revision of Geologic Structure and Stratigraphy." *Colorado Scientific Proceedings*, v. 12, No. 6, p. 163, 1930

___ "Structural Control of Ore Deposition in the Uncompahgre District, Ouray County, Colorado." *U.S. Geological Survey Bulletin* 906-e, pp. 189–265, 1941

___ "Structural Control of Ore Deposits, Red Mountain, Sneffels and Telluride Districts of San Juan Mountains." *Colorado Scientific Society Proceedings*, v. 14, No.5, 1941

Byerly, P. E.; Joesting, H. R. "Regional Geophysical Investigations of the Lisbon Valley Area, Utah and Colorado." U.S. Geological Survey Professional Paper 316-C, pp. 39–50, 1959

Callihan, J. T. "Geology of the Glen Canyon Group Along Echo Cliffs, Arizona." *Plateau*, V. 23, PP. 49–57, 1951

Campbell, J. A.; Baars, D. L. "Environmental and Stratigraphic Significance of Devonian Stromatolites of Colorado" (abstract). *American Association of Petroleum Geologists Bulletin*, v. 48, P. 511, 1964

Case, J. E.; Joesting, H. R.; Byerly, P. E. "Regional Geophysical Investigations in the La Sal Mountains Area, Utah and Colorado." U.S. Geological Survey Professional Paper 316-F, pp. 91–116, 1963

Chenoweth, W. L. "The Uranium Deposits of the Lukachukai Mountains, Arizona." In F. D. Trauger, ed., New Mexico Geological Society, *18th Field Conference Guidebook*, 1967

___ "Uranium in the San Juan Basin—An Overview." In J. E. Fassett and H. L. James, eds., New Mexico Geological Society, *28th Field Conference Guidebook*, 1977

___ "Ambrosia Lake, New Mexico—Giant Uranium District." In O. J. Anderson and others, eds., New Mexico Geological Society, *40th Field Conference Guidebook*, 1989

Chenoweth, W. L.; Malan, R. C. "The Uranium Deposits of Northeastern Arizona." In H. L. James, Ed., New Mexico Geological Society, *24th Field Conference Guidebook*, 1973

Childs, O. E. "Geomorphology of the Valley of the Little Colorado River." *Geological Society of America Bulletin*, v. 59, pp. 353–88, 1948

Chronic, H. "Molluscan Fauna from the Permian Kaibab Formation, Walnut Canyon, Arizona." *Geological Society of America Bulletin*, v. 63, pp. 95–166, 1952

Chuvashov, B. I. "The Carboniferous-Permian Boundary in the USSR." In B. R. Wardlaw, ed., Working Group on the Carboniferous-Permian Boundary. *28th International Geological Congress Proceedings*, pp. 42–56, 1989

Clair, J. R. "Paleozoic Rocks of the Southern Paradox Basin." Four Corners Geological Society Symposium, pp. 36–39, 1952

Coffin, R. C.; Perrini, V. C.; Collins, M. J. "Western Colorado Anticlines." *Colorado Geological Survey Bulletin* 24, 1924

Condon, S. M., "Modifications to Middle and Upper Jurassic Nomenclature in the Southeastern San Juan Basin, New Mexico." In O. J. Anderson, and others eds., New Mexico Geological Society, *40th Field Conference Guidebook*, pp. 231–238, 1989

Cowie, J. W.; Bassett, M. G. "Global Stratigraphic Chart." *International Union of Geological Sciences, Bureau of International Commission on Stratigraphy (ICS:IUGS). Supplement to Episodes*, v. 12, n. 2, Chart.

Craig, L. C.; Cadigan, R. A. "The Morrison and Adjacent Formations in the Four Corners Area." Intermountain Association of Petroleum Geologists, *9th Field Conference Guidebook*, pp. 182–92, 1958

Craig, L. C.; Dickey, D. D. "Jurassic Strata of Southeastern Utah and Southwestern Colorado." Intermountain Association of Petroleum Geologists, *7th Field Conference Guidebook*, pp. 93–104, 1956

Craig, L. C.; Holmes, C. N.; et al. "Preliminary Report on the Stratigraphy of the Morrison and Related Formations of the Colorado Plateau Region." U.S. Geological Survey for Atomic Energy Commission, TEI-180, 1951

Cross, W. "The Laccolithic Mountain Groups of Colorado, Utah and Arizona." *Fourteenth Annual Report of U.S. Geological Survey*, Part II d, pp. 157–241, 1893

___ "Telluride Folio." U.S. Geological Survey Atlas, Folio 57, 1899

___ "A Report on the Economic Geology of the Silverton Quadrangle." *U.S. Geological Survey Bulletin* 182, p. 265, 1901

___ "A New Devonian Formation in Colorado." *American Journal of Science*, 4th Series, v. 18, pp. 245–52, 1904

___ "Stratigraphic Results of a Reconnaissance in Western Colorado and Eastern Utah." *Journal of Geology*, v. 15, p. 666, pp. 634–79, 1907

___ "Geology and Ore Deposits Near Lake City." *U.S. Geological Survey Bulletin* 478, p. 128, 1911

Cross, W.; Hole, A. D. "Engineer Mountain." U.S. Geological Survey Atlas, Folio 171, 1910

Cross, W.; Howe, E. "Red Beds of Southwestern Colorado and Their Correlation." *Geological Society of America Bulletin*, v. 16, p. 476, 1905

Cross, W.; Howe, E.; Irving, J. D. "Ouray Folio." U.S. Geological Survey Atlas, Folio 153, 1907

Cross, W.; Howe, E.; Irving, J. D.; Emmons, W. H. "Needle Mountain Folio." U.S. Geological Survey Atlas, Folio 131, 1905

Cross, W.; Howe, E.; Ransome, F. L. "Silverton Folio." U.S. Geological Survey Atlas, Folio 120, 1905

Cross, W.; Larsen, E. S. "Contributions to the Stratigraphy of Southwestern Colorado." U.S. Geological Survey Professional Paper 90-E, pp. 39–50, 1913

___ "A Brief Review of the Geology of the San Juan Region of Southwestern Colorado." *U.S. Geological Survey Bulletin* 843, p. 138, 1935

___ "Geology and Petrology of the San Juan Mountains, Colorado." *U.S. Geological Survey Bulletin* 843, 1952

Cross, W.; Purington, C. W. "La Plata Folio." U.S. Geological Survey Atlas, Folio 60, 1899

Cross, W.; Spencer, A. C. "Geology of the Rico Mountains, Colorado." *U.S. Geological Survey 21st Annual Report*, Part 2, pp. 7–165, 1900

___ "Rico Folio." *U.S. Geological Survey Atlas*, Folio 130, 1905

Dake, E. L. "The Pre-Moenkopi Unconformity of the Colorado Plateau." *Journal of Geology,* v. 28, No. 1, 1920

Dane, C. H. "Uncompahgre Plateau and Related Structural Features" (abstract). *Washington Academy of Science Journal,* v. 21, No. 2, p. 28, 1931

___ "Geology of Salt Valley Anticline and Adjacent Areas, Grand County, Utah." *U.S. Geological Survey Bulletin 863,* p. 184, 1935

___ "Stratigraphic Relations of Eocene, Paleocene and Latest Cretaceous Formations of Eastern Side of San Juan Basin, New Mexico." U.S. Geological Survey Preliminary Chart No. 24, Oil and Gas Investigation Series, 1946

Darton, N. H. "A Reconnaissance of Parts of Northwestern New Mexico and Northern Arizona." *U.S. Geological Survey Bulletin 435,* p. 88, 1910

___ "Geologic Structures of Parts of New Mexico." *U.S. Geological Survey Bulletin 726-e,* pp. 173–275, 1922

___ "The Permian of Arizona and New Mexico." *American Association of Petroleum Geologists Bulletin,* v. 10, pp. 189–252, 1926

___ "Red Beds and Associated Formations in New Mexico." *U.S. Geological Survey Bulletin 794,* p. 356, 1928

Daugherty, L. H. "The Upper Triassic Flora of Arizona." Carnegie Institute of Washington Publication 526, p. 108, 1941

Davis, G. H. "Monocline Fold Pattern of the Colorado Plateau." *In* V. Matthews, ed., *Geological Society of America Memoir 151,* pp. 215–233, 1978

Davis, L. M. "Characteristics, Occurrence and Uses of the Solid Bitumens of the Uinta Basin, Utah." *The Compass,* v. 29, No. 1, 1951

Davydov, V. I. "The Carboniferous-Permian Boundary in the USSR and its Correlation." Program with Abstracts, *International Congress of the World,* Perm', USSR, p. A3, 1993

Dubiel, R. F. "Sedimentology and Revised Nomenclature for the Upper Part of the Upper Triassic Chinle Formation and the Lower Jurassic Wingate Sandstone, Northwestern New Mexico and Northeastern Arizona." In O. J. Anderson and others, eds., New Mexico Geological Society, *40th Field Conference Guidebook,* pp. 213–223, 1989

Dutton, C. E. "The Physical Geology of the Grand Canyon District." *Second Annual Report of the U.S. Geological Survey,* pp. 47-166, 1881

___ "Tertiary History of the Grand Canyon District." U.S. Geological Survey Monograph 2, p. 264, 1882

___ "Mount Taylor and the Zuni Plateau." U.S. Geological Survey, *6th Annual Report,* pp. 105–98, 1885

Eardley, A. J. "Structural Evolution of Utah." Oil and Gas Possibilities of Utah, Utah Geological and Mineralogical Survey, pp. 10–23, 1949

___ "Paleotectonic and Paleogeologic Maps of Central and Western North America." *American Association Petroleum Geologists Bulletin,* v. 33, pp. 655–82, 1949

Eastman, C. R. "On Upper Devonian Fish Remains from Colorado." *American Journal of Science,* 4th Series, v. 18, pp. 253–60, 1904

Easton, W. H.; Gutschick, R. C. "Corals from the Redwall Limestone (Mississippian) of Arizona." *Bulletin of the Southern California Academy of Sciences,* v. 52, Part 1, pp. 1–27, 1953

Eckel, E. B. "Geology and Ore Deposits of the La Plata District, Colorado." U.S. Geological Survey Professional Paper 219, p. 179, 1949

Edwards, J. Jr. "The Petrology and Structure of the Buried Precambrian Basement of Colorado." *Quarterly of the Colorado School of Mines*, v. 61, No. 4, 1966

Ellingson, J. A. "The Mule Ear Diatreme." In D. L. Baars, ed., *Four Corners Geological Society Guidebook*, 1973

Elston, D. P.; Landis, E. R. "Pre-Cutler Unconformities and Early Growth of the Paradox Valley and Gypsum Valley Salt Anticlines, Colorado." U.S. Geological Survey Professional Paper 400-B, Art. 118, pp. B261-B265, 1960

Elston, D. P.; Shoemaker, E. M. "Late Paleozoic and Early Mesozoic Structural History of the Uncompahgre Front." *Four Corners Geological Society Guidebook, 3rd Field Conference*, pp. 47–55, 1960

Elston, D. P.; Shoemaker, E. M.; Landis, E. R. "Uncompahgre Front and Salt Anticline Region of Paradox Basin, Colorado and Utah." *American Association of Petroleum Geologists Bulletin*, v. 46, No. 10, pp. 1857–78, 1962

Elston, W. E. "Structural Development and Paleozoic Stratigraphy of Black Mesa Basin, Northeastern Arizona, and Surrounding Areas." *American Association of Petroleum Geologists Bulletin*, v. 44, pp. 21–36, 1960

Fassett, J. E.; Hinds, J. S. "Geology and Fuel Resources of the Fruitland Formation and Kirtland Shale of the San Juan Basin." U.S. Geological Survey Professional Paper 676, 1971

Finley, E. A. "Geology of the Dove Creek Area, Dolores and Montezuma Counties, Colorado."U.S. Geological Survey Preliminary Map, OM-120, Oil and Gas Investigation Series, 1951

Fitzsimmons, J. P. "Precambrian of the Four Corners Area." *Four Corners Geological Society Symposium: Shelf Carbonates of the Paradox Basin*, 1963

___ "Tertiary Igneous Rocks of the Navajo Country, New Mexico and Utah." In H. L. James, Ed., New Mexico Geological Society, *24th Field Conference Guidebook*, pp. 106–109, 1973

Foster, R. W. "Stratigraphy of West-Central New Mexico." Four Corners Geological Society, *2nd Field Conference Guidebook*, pp. 62–72, 1957

Gilbert, G. K. "Report on the Geology of Portions of New Mexico and Arizona." *U.S. Geographical and Geological Surveys West of the 100th Meridian Report*, v. 3, *Geology*, pp. 503–67, 1875

Gilluly, J. "Geology and Oil and Gas Prospects of Part of the San Rafael Swell, Utah." *U.S. Geological Survey Bulletin* 806, Part C, pp. 69–130, 1929

Gilluly, J.; Reeside, J. B. Jr. "Sedimentary Rocks of the San Rafael Swell and Some Adjacent Areas in Eastern Utah." U.S. Geological Survey Professional Paper 150, 1928

Ginsburg, R. N.; Isham, L. B.; Bein, S. J.; Kuperberg, G. J. "Laminated Algal Sediments of South Florida and Their Recognition in the Fossil Record." Unpublished report, University of Miami Marine Laboratory, No. 54-20, pp. 1–33, 1954

Ginsburg, R. N.; Lowenstam, H. A. "The Influence of Marine Bottom Communities on the Depositional Environment of Sediments." *Journal of Geology*, v. 66, No. 3, pp. 310–18, 1958

Girty, G. H. "Devonian Fossils from Southwest Colorado; The Fauna of the Ouray Limestone." *Twentieth Annual Report of U.S. Geological Survey*, Part I c, pp. 25–81, 1899

___ "The Carboniferous Formations and Faunas of Colorado." U.S. Geological Survey Professional Paper 16, 1903

Green, M. W.; Pierson, C. T. "A Summary of the Stratigraphy and Depositional Environments of Jurassic and Related rocks in the San Juan Basin, Arizona, Colorado and New Mexico." J. E. Fassett and H. L. James, eds., New Mexico Geological Society, *28th Field Conference Guidebook,* pp. 147–152, 1977

Gregory, H. E. "The Shinarump Conglomerate." *American Journal of Science,* 4th Series, v. 35, pp. 424–38, 1913

___ "The Navajo Country, (Arizona)." *Geological Society of America Bulletin,* v. 47, pp. 561–77, 1915

___ "The Navajo Country, a Geographic and Hydrographic Reconnaissance of Parts of Arizona, New Mexico, and Utah." U.S. Geological Survey Water Supply Paper 380, 1916

___ "Geology of the Navajo Country." U.S. Geological Survey Professional Paper 93, 1917

___ "The San Juan Country." U.S. Geological Survey Professional Paper 188, 1938

___ "Geology and Geography of the Zion Park Region, Utah and Arizona." U.S. Geological Survey Professional Paper 220, 1952

Gregory, H. E.; Moore, R. C. "The Kaiparowits Region, a Geographic and Geologic Reconnaissance of Parts of Utah and Arizona." U.S. Geological Survey Professional Paper 164, 1931

Grundy, W. D.; Oertell, E. W. "Uranium Deposits in the White Canyon and Monument Valley Mining Districts, San Juan County, Utah, and Navajo and Apache Counties, Arizona." Intermountain Association of Petroleum Geologists, *9th Field Conference Guidebook,* pp. 197–207, 1958

Hack, J. T. "Sedimentation and Vulcanism in the Hopi Buttes." *Geological Society of America Bulletin,* v. 53, pp. 335–72, 1942

Hallgarth, W. E. "Stratigraphy of Paleozoic Rocks in Northwestern Colorado." U.S. Geological Survey Oil and Gas Investigations Chart OC-59, 1959

Halpenny, L. C. "Preliminary Report on the Groundwater Resources of the Navajo-Hopi Indian Reservation, Arizona, New Mexico and Utah." *New Mexico Geological Society Guidebook, Second Field Conference,* pp. 147–54, 1951

Hamblin, W. K. "Late Cenozoic Lava Dams in the Western Grand Canyon." In S. S. Beus and M. Morales, eds., *Grand Canyon Geology,* Oxford University Press and Museum of Northern Arizona Press, pp. 385–434, 1990

Harrison, T. S. "Colorado-Utah Salt Domes." *American Association of Petroleum Geologists Bulletin,* v. 11, pp. 111–33, 1927

Harshbarger, J. W. "Upper Jurassic Stratigraphy of Black Mesa, Arizona" (abstract). *Geological Society of America Bulletin,* v. 59, No. 12, Part 2, 1948

___ "The Cow Springs Sandstone of the Black Mesa Basin and Adjoining Areas" (abstract). *American Association of Petroleum Geologists Bulletin,* v. 36, 1952

Harshbarger, J. W.; Repenning, C. A.; Irwin, J. H. "Stratigraphy of the Uppermost Triassic and the Jurassic Rocks of the Navajo Country." New Mexico Geological Society, *9th Field Conference Guidebook,* pp. 98–114, 1958

Harshbarger, J. W.; Repenning, C. A.; Jackson, R. L. "Jurassic Stratigraphy of the Navajo Country-Measured Section Cow Spring Sandstone by J. W. Harshbarger." *New Mexico Geological Society Guidebook, 2nd Field Conference,* pp. 95–99, 1951.

Hayden, F. V. "Geological and Geographical Atlas of Colorado and Portions of Adjacent Territory." U.S. Geographical and Geological Survey, Sheets 11, 14, and I5, 1877

Heaton, R. L. "Ancestral Rockies and Mesozoic and Paleozoic Stratigraphy of Rocky Mountain Region." *American Association of Petroleum Geologists Bulletin*, v. 17, pp. 109–68, 1933

___ "Late Paleozoic and Mesozoic History of Colorado and Adjacent Areas." *American Association of Petroleum Geologists Bulletin*, v. 34, pp. 1659–98, 1950

Herman, G.; Barkell, C. A. "Pennsylvanian Stratigraphy and Productive Zones." *American Association of Petroleum Geologists Bulletin*, v. 41, pp. 861–81, 1957

Herman, G.; Sharps, S. L. "Pennsylvanian and Permian Stratigraphy of the Paradox Salt Embayment." Intermountain Association of Petroleum Geologists, *7th Field Conference Guidebook*, pp. 77–84, 1956

Heylmun, E. B. "Paleozoic Stratigraphy and Oil Possibilities of Kaiparowits Region, Utah." *American Association of Petroleum Geologists Bulletin*, v. 42, pp. 1781–1811, 1958

Hillebran, J. R.; Kelley, V. C. "Mines and Ore Deposits from Red Mountain Pass to Ouray, Ouray County, Colorado." *New Mexico Geological Society Guidebook, 6th Field Conference*, pp. 188–98, 1955

Hinds, N. E. A. "Uncompahgran and Beltian Deposits in Western North America." Carnegie Institute of Washington, Publication No. 463, 1936

___ "Pre-Cambrian Formations in Western North America." *Proceedings of the 6th Pacific Science Congress*, 1939

Hite, R. J. "Stratigraphy of the Saline Facies of the Paradox Member of the Hermosa Formation of Southeastern Utah and Southwestern Colorado." *Four Corners Geological Society Guidebook, 3rd Field Conference*, pp. 86–89, 1960

Hoover, W. B. "Regional Structure of Four Corners Area." *Symposium, Four Corners Geological Society*, pp. 10–11, 1952

Hopkins, R. L. "Kaibab Limestone." In S. S. Beus and M. Morales, eds., *Grand Canyon Geology*, Oxford University Press and Museum of Northern Arizona Press, pp. 225–246, 1990

Huddle, J. W.; Dobrovolny, E. "Late Paleozoic Stratigraphy and Oil and Gas Possibilities of Central and Northeastern Arizona." U.S. Geological Survey Preliminary Chart 10, Oil and Gas Investigations Series, 1945

___ "Devonian and Mississippian Rocks of Central Arizona." U.S. Geological Survey Professional Paper 233-D, pp. 67–112, 1952

Hughes, P. W. "History of the Supai Formation in Black Mesa, Yavapai County, Arizona." *Plateau*, v. 22, pp. 32–36, 1949

___ "Stratigraphy of Supai Formation, Chino Valley Area, Yavapai County, Arizona." *American Association of Petroleum Geologists Bulletin*, v. 36, pp. 635–57, 1952

Hunt, A. P.; Lucas, S. G. "Stratigraphy, Paleontology and Age of the Fruitland and Kirtland Formations, (Upper Cretaceous) San Juan Basin, New Mexico." In S. G. Lucas, and others, eds., New Mexico Geological Society, *43rd Field Conference Guidebook*, pp. 217–240, 1992

Hunt, C. B. "Tertiary Structural History of Part of Northwest New Mexico" (abstract). *Washington Academy of Science Journal*, v. 24, pp. 188–189, 1934

___ "Igneous Geology and Structure of the Mount Taylor Volcanic Field, New Mexico." U.S. Geological Survey Professional Paper 189-B, Part b, pp. 51–80, 1937

Hunt, C. B.; Averitt, P.; Miller, R. L. "Geology and Geography of the Henry Mountains Region, Utah." U.S. Geological Survey Professional Paper 228, 1953

Hunt, C. B.; Dane, C. H. "Map Showing Geologic Structure of the Southern Part of the San Juan Basin, Including Parts of San Juan, McKinley, Sandoval, Valencia, and Bernalillo Counties, New Mexico." U.S. Geological Survey Map OM-158, 1954

Irving, J. D. "Ore Deposits of the Ouray District, Colorado." *U.S. Geological Survey Bulletin 260*, pp. 50–77, 1905

Irwin, J. H.; Stevens, P. R.; Cooley, M. E. "Geology of the Paleozoic Rocks, Navajo and Hopi Indian Reservations, Arizona, New Mexico and Utah." U.S. Geological Survey Professional Paper 521-C, 1971

Ives, Lt. J. C.; Newberry, J. S. *Report upon the Colorado River of the West, Explored in 1857 and 1858.* Washington, D.C., 1861

Jackson, M. P. A.; Schultz-Ela, D. D.; Hudec, M. R.; Watson, I. A.; Porter, M. L. "Structure and Evolution of Upheaval Dome: A Pinched-off Salt Diapir." *Geological Society of America Bulletin*, v. 110, pp. 1547–1573, 1998

Jackson, R. L. "Stratigraphic Relationships of the Supai Formation of Central Arizona." *Plateau*, v. 24, No. 2, pp. 84–91, 1951

___ "Pennsylvanian-Permian Facies of the Supai Formation in Central Arizona." *Guidebook for Field Trip Excursions in Southern Arizona, Arizona Geological Society*, 1952

Jentgen, R. W. "Pennsylvanian Rocks in the San Juan Basin. New Mexico and Colorado." In J. E. Fassett and H. L. James, eds., New Mexico Geological Society, *28th Field Conference Guidebook*, pp. 129–132, 1977

Joesting, H. R.; Byerly, P. E. "Regional Geophysical Investigations of the Uravan Area, Colorado." U.S. Geological Survey Professional Paper 316-A, pp. 1–17, 1958

Joesting, H. R.; Case, J. E. "Salt Anticlines and Deep-Seated Structures in the Paradox Basin, Colorado and Utah." U.S. Geological Survey Professional Paper 400-B, Article 114, pp. B252-B256, 1960

___ "Regional Geophysical Studies in Salt Valley-Cisco Area, Utah and Colorado." *American Association of Petroleum Geologists Bulletin*, v. 46, pp. 1879–89, 1962

Johnson, J. H. "A Resume of Paleozoic Stratigraphy of Colorado." *Quarterly of the Colorado School of Mines*, v. 40, No. 3, 1945

Jones, D. J.; Picard, M. D.; Wyeth, J. C. "Correlation of Non-Marine Cenozoic of Utah." *American Association of Petroleum Geologists Bulletin*, v. 38, pp. 2219–21, 1954

Jones, R. W. "Origin of Salt Anticlines of Paradox Basin." *American Association of Petroleum Geologists Bulletin*, v. 43, pp. 1869–95, 1959

Kelley, V. C. "Pre-Cambrian Rocks of the San Juan Basin." *New Mexico Geological Society Guidebook, 1st Field Conference*, pp. 53–55, 1950

___ "Regional Structure of the San Juan Basin." *New Mexico Geological Society Guidebook, 1st Field Conference*, pp. 101–8, 1950

___ "Tectonics of the San Juan Basin." *New Mexico Geological Society Guidebook, 2nd Field Conference*, pp. 124–31, 1951

___ "Vein and Fault Systems of the Western San Juan Mountains Mineral Belt, Colorado." *New Mexico Geological Society Guidebook of Southwestern San Juan Mountains, Colorado*, pp. 173–75, 1955

___ "Regional Tectonics of the Colorado Plateau and Relationship to the Origin and Distribution of Uranium." University of New Mexico (Publication in Geology), No. 5, 1955

___ "Tectonics of the Region of the Paradox Basin." Intermountain Association of Petroleum Geologists, *9th Field Conference Guidebook*, 1958

Keyes, C. R. "Carboniferous Formations of New Mexico." *Journal of Geology*, v. 14, pp. 147–54, 1906

Kindle, E. M. "The Devonian Fauna of the Ouray Limestone." *U.S. Geological Survey Bulletin* 391, 1909

Kirk, E. "The Devonian of Colorado." *American Journal of Science*, 5th Series, v. 22, pp. 220–40, 1931

Knight, R. L.; Baars, D. L. "New Developments on Age and Extent of Ouray Limestone." *American Association of Petroleum Geologists Bulletin*, v. 41, pp. 2275–83, 1957

Knight, R. L.; Cooper, J. C. "Suggested Changes in Devonian Terminology of the Four Corners Area." Four Corners Geological Society, *1st Field Conference Guidebook*, pp. 56–58, 1955

Knowlton, F. H. "Contributions to the Geology and Paleontology of San Juan County, New Mexico, 4. Flora of the Fruitland and Kirtland Formations." U.S. Geological Survey Professional Paper 98s, pp. 327–53, 1916

Knowlton, F. H.; Reeside, J. B. "Flora of the Animas Formation. Upper Cretaceous and Tertiary Formations of the Western Part of the San Juan Basin, Colorado and New Mexico." U.S. Geological Survey Professional Paper 134, 1924

Kriens, B. J.; Shoemaker, E. M.; Herkenhoff, K. E. "Geology of the Upheaval Dome Impact Structure, Southeast Utah." *Journal of Geophysical Research*, v. 104, n. E8, pp. 18, 867–18, 887, 1999.

Kunkel, R. P. "Permian Stratigraphy of the Paradox Basin." Intermountain Association of Petroleum Geologists, *9th Field Conference Guidebook*, pp. 163–68, 1958

___ "Permian Stratigraphy in the Salt Anticline Region of Western Colorado and Eastern Utah." Four Corners Geological Society, *3rd Field Conference Guidebook*, pp. 91–97, 1960

Lance, J. F. "Precambrian Rocks of Northern Arizona." New Mexico Geological Society, *9th Field Conference Guidebook*, pp. 66–70, 1958

Lee, W. T. "Relation of the Cretaceous Formations of the Rocky Mountains in Colorado and New Mexico." U.S. Geological Survey Professional Paper 95c, pp. 27–58, 1916

___ "General Stratigraphic Break Between the Pennsylvanian and Permian in Western America." *Geological Society of America Bulletin*, v. 28, pp. 169–70, 1917

Lessentine, R. H. "Kaiparowits and Black Mesa Basins: Stratigraphic Synthesis." *American Association of Petroleum Geologists Bulletin*, v. 49, pp. 1997–2019, 1965

Likharev, B. K. "The Boundaries and Principal Subdivisions of the Permian System," *Soviet Geology*, v. 66, pp. 12–30

Loliet, A. J. "Cambrian Stratigraphic Problems of the Four Corners Area. "Four Corners Geological Society Symposium: Shelf Carbonates of the Paradox Basin, 1963

Lovering, T. S. "Geology and Ore Deposits of the Montezuma Quadrangle, Colorado." U.S. Geological Survey Professional Paper 178, 1935

Lucas, S. G. "Vertebrate Paleontology of the San Jose Formation, East-central San Juan Basin, New Mexico." In J. E. Fassett, ed., *New Mexico Geological Society Guidebook*, pp. 221–25, 1977

Lucas, S. G.; Hayden, S. N. "Triassic Stratigraphy of West-central New Mexico." O. J. Anderson and others, eds., New Mexico Geological Society, *40th Field Conference Guidebook*, pp. 191-211, 1989

Lucchitta, I. "History of the Grand Canyon and of the Colorado River in Arizona." In S. S. Beus and M. Morales, eds., *Grand Canyon*, Oxford University Press and Museum of Northern Arizona Press, pp. 311–32, 1990

Luedke, R. G.; Burbank, W. S. "Geology of the Ouray Quadrangle, Colorado." U.S. Geological Survey, Map GQ-152, 1962

Mallory, W. W. "Pennsylvanian Arkose and the Ancestral Rocky Mountains." In W. W. Mallory, ed., *Geologic Atlas of the Rocky Mountain Region*, Rocky Mountain Association of Geologists, pp. 131–2, 1972

McKee, E. D. "The Coconino Sandstone—Its History and Origin." Carnegie Institute of Washington, Publication No. 440, pp. 77–115, 1933

___ "Investigations of Lightcolored Crossbedded Sandstone of Canyon de Chelly, Arizona." *American Journal of Science*, 5th Series, v. 28, 1934

___ "The Environment and History of the Toroweap and Kaibab Formations of Northern Arizona and Southern Utah." Carnegie Institute of Washington, Publication 492, 1938

___ "Three Types of Cross-Lamination in Paleozoic Rocks of Northern Arizona." *American Journal of Science*, v. 238, pp. 811–24, 1940

___ "Cambrian History of the Grand Canyon Region." Carnegie Institute of Washington, Publication 563, 1945

___ "Facies Changes in the Colorado Plateau." *Geological Society of America Memoir 39*, pp. 35–48, 1949

___ "Sedimentary Basins of Arizona and Adjoining Areas." *Geological Society of America Bulletin*, v. 62., pp. 481–505, 1951

___ "Uppermost Paleozoic Strata of Northwestern Arizona and Southwestern Utah." *Intermountain Association of Petroleum Geologists Guidebook to the Geology of Utah*, No. 7, pp. 52–55, 1952

___ "Stratigraphy and History of the Moenkopi Formation of Triassic Age." *Geological Society of America Memoir 61*, 1954

___ "Primary Structures in Some Recent Sediments." *American Association of Petroleum Geologists*, v. 41, pp. 1704–47, 1957

___ "Lithologic Subdivisions of the Redwall Limestone in Northern Arizona-Their Paleogeographic and Economic Significance." U.S. Geological Survey Professional Paper 400-B, Article 110, pp. B243-B245, 1960

___ "Nomenclature for Lithologic Subdivisions of the Mississippian Redwall Limestone, Arizona." U.S. Geological Survey Professional Paper 475-C, Article 65, pp. C21-C22, 1963

McKee, E. D.; Hamblin, W. K.; Damon, E. "K-Ar Age of Lava Dam in Grand Canyon." Geological *Society of America Bulletin*, v. 79, pp. 133–36, 1968

McKee, E. D.; Wilson, R. F.; Breed, W. J. *Evolution of the Colorado River in Arizona*. Museum of Northern Arizona, Flagstaff, 1964

McKnight, E. T. "Geology of Area Between Green and Colorado Rivers, Grand and San Juan Counties, Utah." *U.S. Geological Survey Bulletin 908*, pp. 1–147, 1940

McLemore, V. T.; Chenoweth, W. L. "Uranium Deposits in the Eastern San Juan Basin, Cibola, Sandoval and Rio Arriba Counties, New Mexico." In S. G. Lucas and others, eds., New Mexico Geological Society, *43rd Field Conference Guidebook*, pp. 341–50, 1992

Melton, F. A. "The Ancestral Rocky Mountains of Colorado and New Mexico." *Journal of Geology*, v. 33, pp. 84–89, 1925

Merrill, W. M.; Winar, R. M. "Molas and Associated Formations in San Juan Basin-Needle Mountains Area, Southwestern Colorado." *American Association Petroleum Geologists Bulletin*, v. 42, pp. 2107–32, 1958

Miser, H. D. "Geologic Structure of the San Juan Canyon and Adjacent Country, Utah." *U.S. Geological Survey Bulletin* 751-D, pp. 115–55, 1924

___ "The San Juan Canyon, Southeastern Utah." U.S. Geological Survey Water Supply Paper 538, 1924

___ "Erosion in San Juan Canyon, Utah." *Geological Society of America Bulletin*, v. 36, pp. 365–77, 1925

Molenaar, C. M. "Stratigraphy and Depositional History of Upper Cretaceous Rocks of the San Juan Basin Area, New Mexico and Colorado, with a note on Economic Resources." In J. E. Fassett and H. L. James, eds., New Mexico Geological Society, *28th Field Conference Guidebook*, pp. 159–166, 1977

___ "Major Depositional Cycles and Regional Correlations of Upper Cretaceous Rocks, Southern Colorado Plateau and Adjacent Areas." In M. W. Reynolds and E. D. Dolly, eds., *Mesozoic Paleogeography of West-central United States, Rocky Mountain Section, Society of Economic Paleontologists and Mineralogists*, pp. 201–224, 1983

Molenaar, C. M.; Rice, E. D. "Cretaceous Rocks of the Western Interior Basin." In L. L. Sloss, ed., *Sedimentary Cover, North American Craton, U.S.*, Geological Society of America, The Geology of North America, v. D-2, pp. 77–82, 1988

Neff, A. W.; Brown, S. C. "Ordovician-Mississippian Rocks of the Paradox Basin." Intermountain Association of Petroleum Geologists, *9th Field Conference Guidebook*, pp. 102–8, 1958

Newberry, J. S. "Geological Report in Report of the Exploring Expedition from Santa Fe, New Mexico, to the Junction of the Grand and Green Rivers of the Great Colorado of the West in 1859. Under the Command of Capt. J. N. Macomb." U.S. Army Engineer Department. Pp. 135–48, 1876

Noble, L. F. "The Shinumo Quadrangle, Grand Canyon District, Arizona." *U.S. Geological Survey Bulletin* 549, 1914

___ "A Section of the Paleozoic Formations of the Grand Canyon at the Bass Trail." U.S. Geological Survey Professional Paper 131, Part b, pp. 23–73. 1923

___ "A Section of the Kaibab Limestone in Kaibab Gulch, Utah." U.S. Geological Survey Professional Paper 150, Part c, pp. 41–60, 1928

Noble, L. F.; Hunter, J. F. "A Reconnaissance of the Archean Complex of Granite Gorge, Grand Canyon, Arizona." U.S. Geological Survey Professional Paper 98, Part 1, pp. 53–113, 1917

Northrop, S. A.; Wood, G. H. "Geology of Nacimiento Mountains, San Pedro Mountains and Adjacent Plateaus in Part of Sandoval and Rio Arriba Counties, New Mexico." U.S. Geological Survey Preliminary Map 57, Oil and Gas Investigations Series, 1946

O'Sullivan, R. B. "Triassic Rocks in the San Juan Basin of New Mexico and Adjacent Areas." In J. E. Fassett and H. L. James, eds., *New Mexico Geological Society 28th Field Conference Guidebook*, pp. 139–146, 1977

O'Sullivan, R. B.; Green, M. W. "Triassic Rocks of Northeast Arizona and Adjacent Areas." In H. L. James, ed., New Mexico Geological Society, *24th Field Conference Guidebook*, pp. 72–78, 1973

O'Sullivan, R. B.; Craig, L. C. "Jurassic Rocks of Northeast Arizona and Adjacent Areas." In H. L. James, ed., New Mexico Geological Society, *24th Field Conference Guidebook*, pp. 79–85, 1973

Page, H. G.; Repenning, C. A. "Late Cretaceous Stratigraphy of Black Mesa, Navajo and Hopi Indian Reservations, Arizona." New Mexico Geological Society, *9th Field Conference Guidebook*, pp. 115–22, 1958

Parker, J. M. "The McIntyre Canyon and Lisbon Oil and Gas Fields, San Miguel County, Colorado, and San Juan County, Utah." *Rocky Mountain Association of Geologists, Symposium on Lower and Middle Paleozoic Rocks of Colorado,* pp. 163–73, 1961

___ "The Cambrian, Devonian, and Mississippian Rocks and Pre-Pennsylvanian Structure of Southwest Colorado and Adjoining Portions of Utah, Arizona, and New Mexico." *Rocky Mountain Association of Geologists, Symposium on Lower and Middle Paleozoic Rocks of Colorado,* pp. 59–70, 1961

___ "Pre-Pennsylvanian Beds Yield Lion's Share of Oil in Lisbon Area." *Oil and Gas Journal,* April 16, 1962

Parker, J. W. "Big Flat Field, Utah." *Four Corners Geological Society Guidebook, 3rd Field Conference,* pp. 127–32, 1960

Parker, J. W.; Roberts, J. W. "Regional Devonian and Mississippian Stratigraphy, Central Colorado Plateau." *American Association of Petroleum Geologists Bulletin,* v. 50, pp. 2404–33, 1966

Peabody, F. "Reptile and Amphibian Trackways from Lower Triassic Moenkopi of Arizona and Utah." *University of California Publications Bulletin, Department of Geological Science,* v. 27, pp. 195–468, 1948

Peterson, F. "A Synthesis of the Jurassic System in the Southern Rocky Mountain Region." In L. L. Sloss, ed., *Sedimentary Cover, North American Craton, U.S.,* Geological Society of America, The Geology of North America, v. D-2, pp. 65–76, 1988

Peterson, F.; Kirk, A. R. "Correlation of Cretaceous Rocks in the San Juan, Kaiparowits and Henry Basins, Southern Colorado Plateau." In J. E. Fassett and H. L. James, eds., New Mexico Geological Society, *28th Field Conference Guidebook,* pp. 167–178, 1977

Peterson, F.; Pipiringos, G. N. "Stratigraphic Relationships of the Navajo Sandstone to Middle Jurassic Formations, Southern Utah and Northern Arizona." U.S. Geological Survey Professional Paper 1035-B

Peterson, J. A. "Petroleum Geology of the Four Corners Area." Fifth World Petroleum Congress, Paper 27, Section 1, 1959

Peterson, J. A.; Ohlen, H. R. "Pennsylvania Shelf Carbonates, Paradox Basin." *Four Corners Geological Society Symposium: Shelf Carbonates of the Paradox Basin,* pp. 65-79, 1963

Pierce, H. W. "Permian Sedimentary Rocks of the Black Mesa Basin Area." New Mexico Geological Society, *9th Field Conference Guidebook,* pp. 82–87, 1958

___ "Permian Stratigraphy of the Defiance Plateau." In E. D. Trauger, ed., New Mexico Geological Society, *18th Field Conference Guidebook,* pp. 57–62, 1967

Pike, W. S. Jr. "Intertonguing Marine and Non-Marine Upper Cretaceous Deposits of New Mexico, Arizona and Southwestern Colorado." *Geological Society of America Memoir 24,* 1947

Poole, F. G.; Baars, D. L.; Drewes, H.; Hayes, P. T.; Ketner, K. B.; McKee E. D.; Teichert, C.; Williams, J. S. "Devonian System of Southwestern United States." Alberta Society of Petroleum Geologists, International Symposium on the Devonian System, 1967

Powell, J. W. *Exploration of the Colorado River of the West and Its Tributaries.* U.S. Government Printing Office, Washington, D.C., pp. 3–145, 1875

Powers, S. "Effect of Salt and Gypsum on tile Formation of Paradox and Other Valleys of Southwestern Colorado." (abstract). *Geological Society of America Bulletin,* v. 37, 1926

Pray, L. C.; Wray, J. L. "Porous Algal Facies (Pennsylvanian) Honaker Trail, San Juan County, Utah." *Four Corners Geological Society Symposium: Shelf Carbonates of the Paradox Basin,* pp. 204–34, 1963

Prommel, H. W. C.; Crum, H. E. "Geology and Structure of Portions of Grand and San Juan Counties, Utah." *American Association of Petroleum Geologists Bulletin,* v. 7, pp. 384–99, 1923

___ "Salt Domes of Permian and Pennsylvanian Age in Southeastern Utah." *American Association of Petroleum Geologists Bulletin,* v. II, pp. 373–393, 1927

___ "Structural History of Parts of Southeastern Utah from Interpretation of Geologic Sections." *American Association of Petroleum Geologists Bulletin,* v. II, pp. 809–20, 1927

Ransome, F. L. "A Report on the Economic Geology of the Silverton Quadrangle, Colorado." *U.S. Geological Survey Bulletin* 182, 1901

___ "Ore Deposits of the Rico Mountains, Colorado." *U.S. Geological Survey, 22nd Annual Report,* Part 2, pp. 229–398, 1901

___ "Some Paleozoic Sections in Arizona and Their Correlation." U.S. Geological Survey, Professional Paper 98-K, pp. 133–66, 1917

Read, C. B. "Stratigraphy of Outcropping Permian Rocks Around the San Juan Basin." *New Mexico Geological Guidebook, 1st Field Conference,* pp. 62–66, 1950

Read, C. B.; Wood, G. H. "Distribution and Correlation of Pennsylvanian Rocks in Late Paleozoic Sedimentary Basins of Northern New Mexico." *Journal of Geology,* v. 55, pp. 220–36, 1947

Reagan, A. B. "Stratigraphy of the Hopi Buttes Volcanic Field." *Pan American Geologist,* v. 41, pp. 355–66, 1924

___ "Late Cretacic Formations of Black Mesa, Arizona." *Pan American Geologist,* v. 44, pp. 285–94, 1925

Reeside, J. B. Jr. "Triassic-Jurassic 'Red Beds' of the Rocky Mountain Region." *Journal of Geology,* v. 37, No. 1, 1929

Reeside J. B. Jr.; Baker, A. A. "The Cretaceous Section in Black Mesa, Northern Arizona." *Washington Academy of Science Journal,* v. 19, pp. 30–37, 1929

Reeside, J. B. Jr.; Hunt, C. B.; Hendricks, T. A. "Transgressive and Regressive Cretaceous Deposits in Southern San Juan Basin, New Mexico." U.S. Geological Survey Professional Paper 193, pp. 101–21, 1941

Reeside, J. B. Jr.; Knowlton, F. H. "Upper Cretaceous and Tertiary Formations of the Western Part of the San Juan Basin, Colorado and New Mexico." U.S. Geological Survey Professional Paper 134, 1924

Reiche, P. "An Analysis of Cross-Lamination: The Coconino Sandstone." *Journal of Geology,* v. 49, No. 7, pp. 905–32, 1938

Repenning, C. A.; Irwin, J. H. "Bidahochi Formation of Arizona and New Mexico." *American Association of Petroleum Geologists Bulletin,* v. 38, pp. 1821–26, 1954

Repenning, C. A.; Lance, J. F.; Irwin, J. H. "Tertiary Stratigraphy of the Navajo Country." New Mexico Geological Society, *9th Field Conference Guidebook,* pp. 123–29, 1958

Roebuck, R. C. "Chinle and Moenkopi Formations, Southeastern Utah." Intermountain Association of Petroleum Geologists, *9th Field Conference Guidebook,* pp. 169–71, 1958

Robinson, H. H. "The San Franciscan Volcanic Field, Arizona." U.S. Geological Survey Professional Paper 76, 1913

Ross, C. A.; Ross, J. R. P. "The Need for a Bursumian Stage, Uppermost Carboniferous, North America." *Permophiles*, v. 24, pp. 3–6, 1994

___ "Bursumian Stage, Uppermost Carboniferous of Midcontinent and Southwestern North America." *Carboniferous Newsletter*, v. 6, pp. 40–42, 1998

Roth, R. I. "Type Section of the Hermosa Formation, Colorado." *American Association of Petroleum Geologists Bulletin*, v. 18, pp. 944–47, 1934

Sando, W. J. "Stratigraphic Importance of Corals in the Redwall Limestone, Northern Arizona." U.S. Geological Survey Professional Paper 501-C, pp. C39-C42, 1964

Schneider, H. "Summary of the Cambrian Stratigraphy of Utah." *Oil and Gas Possibilities of Utah*, Utah Geological and Mineralogical Survey. pp. 31–37, 1949

Schuchert, C. "On the Carboniferous of the Grand Canyon of Arizona." *American Journal of Science*, 4th Series, v. 45, pp. 347–69, 1918

Sears, J. D. "Geology and Coal Resources of the Gallup-Zuni Basin, New Mexico." *U.S. Geological Survey Bulletin 767*, 1925

___ "Geology of Comb Ridge and Vicinity North of San Juan River, San Juan County, Utah." *U.S. Geological Survey Bulletin 1021-E*, 1956

Sears, J. D.; Hunt, C. B.; Hendricks, T. A. "Transgressive and Regressive Cretaceous Deposits in Southern San Juan Basin, New Mexico." U.S. Geological Survey Professional Paper 193-F, pp. 101–21, 1941

Shoemaker, E. M. "Structural Features of Southeastern Utah and Adjacent Parts of Colorado, New Mexico, and Arizona." Utah Geological Society, *Guidebook to the Geology of Utah*, No. 9, pp. 48–69, 1954

Shoemaker, E. M.; Case, J. E.; Elston, D. P. "Salt Anticlines of the Paradox Basin." Intermountain Association of Petroleum Geologists, *9th Field Conference Guidebook*, pp. 39–59, 1958

Silver, C. "Cretaceous Stratigraphy of the San Juan Basin." *New Mexico Geological Society Guidebook, 2nd Field Conference*, pp. 104–17, 1951

Simpson, G. G. "The Eocene of the San Juan Basin, New Mexico." *American Journal of Science*, v. 246, pp. 257–82 (Part 1), pp. 365–85 (Part II), 1948

___ "Lower Tertiary Formations and Vertebrate Faunas of the San Juan Basin." *New Mexico Geological Society Guidebook, 1st Field Conference*, pp. 85–90, 1950

Sinclair, W. J.; Granger, W. "Paleocene Deposits of the San Juan Basin, New Mexico." *American Museum of Natural History Bulletin*, v. 33, pp. 297–316, 1914

Smith, C. T. "Geology of the Thoreau Quadrangle, Valencia and McKinley Counties, New Mexico." *New Mexico Bureau of Mines Bulletin 3*, 1954

Spencer, A. C. "Devonian Strata in Colorado." *American Journal of Science*, 4th Series, v. 9, pp. 125–33, 1900

Spieker, E. M. "Late Mesozoic and Early Cenozoic History of Central Utah." U.S. Geological Survey Professional Paper 205-D, pp. 117–61, 1946

___ "Sedimentary Facies and Associated Diastrophism in the Upper Cretaceous of Central and Eastern Utah." *Geological Society of America Memoir 39*, pp. 55–82, 1949

___ "The Transition Between the Colorado Plateaus and the Great Basin in Central Utah." Utah Geological Society, *Guidebook to the Geology of Utah*, No. 4, 1949

Stevenson, J. J. "Report upon Geological Examinations in Southern Colorado and Northern New Mexico During the Years 1878 and 1879." *U.S. Geological and Geographical Surveys West of the 100th Meridian Report*, v. 3, 1881

Stewart, J. H. "Triassic Strata of Southeastern Utah and Southwestern Colorado." Intermountain Association of Petroleum Geologists, *7th Field Conference Guidebook*, pp. 85–92, 1956

___ "Stratigraphic Relations of Hoskinnini Member (Triassic?) of Moenkopi Formation on Colorado Plateau." *American Association of Petroleum Geologists Bulletin*, v. 43, pp. 1852–68, 1959

Stewart, J. H.; Williams, G. A.; Albee, H. F.; Ralip, O. B.; Cadigan, R. A. "Stratigraphy of Triassic and Associated Formations in Part of the Colorado Plateau." *U.S. Geological Survey Bulletin* 1046, 1959

Stewart, J. H.; Wilson, R. F. "Triassic Strata of the Salt Anticline Region, Utah and Colorado." Four Corners Geological Society *3rd Field Conference Guidebook*, pp. 98–106, 1960

Stieff, L. R.; Stern, T. W.; Milkey, R. G. "A Preliminary Determination of the Age of Some Uranium Ores of the Colorado Plateaus by the Lead-Uranium Method." U.S. Geological Survey Circular 271, 1953

Stokes, W. L. "Morrison Formation and Related Deposits in and Adjacent to the Colorado Plateau." *Geological Society of America Bulletin*, v. 55, pp. 951–92, 1944

___ "Geology of the Utah-Colorado Salt Dome Region with Emphasis on Gypsum Valley, Colorado." Utah Geological Society, *Guidebook to the Geology of Utah*, No. 3, 1948

___ "Triassic and Jurassic Rocks of Utah." *The Oil and Gas Possibilities of Utah*, Utah Geological and Mineralogical Survey, pp. 78–89, 1949

___ "Pediment Concept Applied to Shinarump and Similar Conglomerates." *Geological Society of America Bulletin*, v. 61, pp. 68–91, 1950

___ "Carnotite Deposits in the Carrizo Mountain Area, Apache County, Arizona, and San Juan County, New Mexico." U.S. Geological Survey Circular 111, 1951

___ "Lower Cretaceous in Colorado Plateau." *American Association of Petroleum Geologists Bulletin*, v. 36, pp. 1766–76, 1952

___ "Uranium-Vanadium Deposits of the Thompsons Area, Grand County, Utah." *Utah Geological and Mineralogical Survey Bulletin* 46, 1952

___ "Continental Sediments of the Colorado Plateau." Intermountain Association of Petroleum Geologists, *9th Field Conference Guidebook*, pp. 26–30, 1958

Stokes, W. L.; Heylmun, E. B. "Outline of the Geologic History and Stratigraphy of Utah." *Utah Geological and Mineralogical Survey*, 1958

Stokes, W. L.; Phoenix, D. A. "Geology of the Egnar-Gypsum Valley Area, San Miguel and Montrose Counties, Colorado." U.S. Geological Survey Preliminary Map 93, Oil and Gas Investigations Series, 1948

Stoyanow, A. A. "Correlation of Arizona Paleozoic Formations." *Geological Society of America Bulletin*, v. 47, pp. 459–540, 1936

___ "Paleozoic Paleogeography of Arizona." *Geological Society of America Bulletin*, v. 53, pp. 1255–82, 1942

Strahler, A. N. "Guide to the East Kaibab Monocline in the Grand Canyon Region." *Plateau*, v. 17, pp. 1–13, 1944

___ "Geomorphology and Structure of the West Kaibab Fault Zone and Kaibab Plateau, Arizona." *Geological Society of America Bulletin*, v.59, pp. 513–40, 1948

Thomas, W. A.; Baars, D. L. "The Paradox Transcontinental Fault Zone." *Oklahoma Geological Survey Circular* 97, pp. 3–12, 1995

Thompson, M. L. "Pennsylvanian System in New Mexico." *New Mexico Bureau of Mines and Mineral Resources Bulletin* 17, 1942

Tweto, O. "Tectonic History of Colorado." *Colorado Geology,* eds. Kent and Porter, Rocky Mountain Association of Geologists, pp. 5–9, 1980.

Umbach, P. H. "Cretaceous Rocks of the San Juan Basin Area." *New Mexico Geological Society Guidebook, 1st Field Conference,* pp. 82–83, 1950

Van Gundy, C. E. "Faulting in East Part of Grand Canyon of Arizona." *American Association of Petroleum Geologists Bulletin,* v. 30, pp. 1899–1909, 1946

Vaughn, P. P. "Vertebrates from the Cutler Group of Monument Valley and Vicinity." In H. L. James, ed., New Mexico Geological Society, *24th Field Conference Guidebook,* pp. 99–105, 1973

Ver Wiebe, W. A. "Ancestral Rocky Mountains." *American Association of Petroleum Geologists Bulletin,* v. 14, pp. 765–88, 1930

Walcott, C. D. "The Permian and Other Paleozoic Groups of the Kanab Valley, Arizona." *American Journal of Science,* 3d Series, v. 20, pp. 221–25, 1880

___ "Pre-Carboniferous Strata in the Grand Canyon of the Colorado, Arizona." *American Journal of Science,* 3d Series, v. 26, 1883

___ "Algonkian Rocks of the Grand Canyon of the Colorado. "*Journal of Geology,* v. 3, pp. 312–30, 1895

Wanek, A. A. "Geologic Map of the Mesa Verde Area, Montezuma County, Colorado." U.S. Geological Survey Map OM-152, 1954

Ward, L. F. "Geology of the Little Colorado Valley." *American Journal of Science,* 4th Series, v. 12, pp. 401–13, 1901

Warner, L. A., "The Colorado Lineament: A Middle Precambrian Wrench Fault System." *Bulletin of the Geological Society of America,* v. 89, pp. 161-71, 1978.

Welles, S. P. "Vertebrates from the Upper Moenkopi Formation of Northern Arizona." *University of California Publications Bulletin, Department of Geological Science,* v. 27, pp. 241–97, 1947

Welsh, J. E. "Faunizones of the Pennsylvanian and Permian Rocks in the Paradox Basin." Intermountain Association of Petroleum Geologists, *9th Field Conference Guidebook,* pp. 153–62, 1958

Wengerd, S. A. "Reef Limestones of Hermosa Formation, Utah." *American Association of Petroleum Geologists Bulletin,* v. 35, pp. 1038–51, 1951

___ "Biohermal Trends in Pennsylvanian Strata of San Juan Canyon, Utah." Four Corners Geological Society, *1st Field Conference Guidebook,* pp. 70–77, 1955

___ "Permo-Pennsylvanian Stratigraphy of the Western San Juan Mountains, Colorado." New Mexico Geological Society, *8th Field Conference Guidebook,* pp. 131–37, 1957

___ "Pennsylvanian Stratigraphy, Southwest Shelf, Paradox Basin." Intermountain Association of Petroleum Geologists, *9th Field Conference Guidebook,* pp. 109–34, 1958

___ "Pennsylvanian Sedimentation in Paradox Basin, Four Corners Region." In C. C. Branson, ed., Pennsylvanian System in the United States, *American Association of Petroleum Geologists, a Symposium,* pp. 264–330, 1962

___ "Regional Stratigraphic Control of the Search For Pennsylvanian Petroleum, Southern Monument Upwarp, Southeastern Utah." In H. L. James, ed., New Mexico Geological Society, *24th Field Conference Guidebook,* pp. 122–138, 1973

Wengerd, S. A.; Matheny, M. L. "Pennsylvanian System of Four Corners Region." *American Association of Petroleum Geologists Bulletin,* v. 42, pp. 2048–2106, 1958

Wengerd, S. A.; Strickland, J. W. "Pennsylvanian Stratigraphy of Paradox Salt Basin, Four Corners Region, Colorado and Utah." *American Association of Petroleum Geologists Bulletin,* v. 38, pp. 2157–99, 1954

Wheeler, R. B.; Kerr, A. R. "Preliminary Report on the Tonto Group of the Grand Canyon, Arizona." Grand Canyon Natural History Association, *Bulletin* 5, 1936

White, D. *Flora of the Hermit Shale, Grand Canyon, Arizona.* Carnegie Institute of Washington Publication 405, pp. 1–221, 1929

Williston, S. W. "A new Family of Reptiles from the Permian of New Mexico." *American Journal of Science,* 4th Series, v. 31, pp. 379–80, 1911

Williston, S. W.; Case, E. C. "The Permo-Carboniferous of Northern New Mexico." *Journal of Geology,* v. 20, pp. 1–12, 1912

Winters, S. S. "Permian Stratigraphy in Eastern Arizona." *Plateau,* v. 24, pp. 10–16, 1951

Wood, G. H.; Northrop, S. A. "Geology of Nacimiento Mountain, San Pedro Mountain and Adjacent Plateaus in Parts of Sandoval and Rio Arriba Counties, New Mexico." U.S. Geological Survey Preliminary Map 57, Oil and Gas Investigations Series, 1946

Wood, H. B.; Lekas, M. A. "Uranium Deposits of the Uravan Mineral Belt." Intermountain Association of Petroleum Geologists, *9th Field Conference Guidebook,* pp. 208–15, 1958

Woodruff, E. G. "Geology of the San Juan Oil Field, Utah." *U.S. Geological Survey Bulletin 471,* pp. 76–104, 1910

Woodward, L. A. "Structural Framework and Tectonic Evolution of the Four Corners Region of the Colorado Plateau." In H. L. James, ed., New Mexico Geological Society, *24th Field Conference Guidebook,* pp. 94–98, 1973

Woodward, L. A.; Callender, J. F. "Tectonic Framework of the San Juan Basin." In J. E. Fassett and H. L. James, eds., New Mexico Geological Society, *28th Field Conference Guidebook,* pp. 209–212, 1977

Wright, J. C.; Dickey, D. D. "Pre-Morrison Jurassic Strata of Southeastern Utah." Intermountain Association of Petroleum Geologists, *9th Field Conference Guidebook,* pp. 172–81, 1958

Young, R. G. "Cretaceous Stratigraphy of the Four Corners Area." In H. L. James, ed., *New Mexico Geological Society 24th Field Conference Guidebook,* pp. 86–93, 1973

Zapp, A. D. "Geology and Coal Resources of the Durango Area, La Plata and Montezuma Counties, Colorado." U.S. Geological Survey Preliminary Map 109, Oil and Gas Investigations Series, 1949

INDEX

Bureau of Reclamation, 37
Bursum Formation, 72, 106

Calcareous algae, 60, 95, 129
Calcium carbonate, 17, 18, 60, 226
Cambrian Period, 4, 10–12
Canyonlands National Park, 45, 54, 63, 67, 69, 71, 78, 79, 92, 136
Canyonlands region, 48, 70, 80, 87, 95, 216
Capitol Reef National Park, 146, 188
Carlsbad Caverns, 111
Carmel Formation, 148, 150–52
Carnotite, 139
Cascade Creek, 121, 125
Cataract Canyon, vii, 31, 45, 54, 56, 59, 67, 70, 72, 92, 96, 99, 195, 210, 216
Cedar Mesa Sandstone, 34, 69, 72, 78, 79, 99, 100, 225
Cedar Mountain Formation, 161
Cephalopods, 25, 137, 163, 166, 224
Chalcopyrite, 190
Chert, 9, 16, 18, 50, 224
Chinle Formation, 32, 101, 137, 140, 141, 143, 219
Chuska Mountains, 220
Coal Bank Pass, 115, 117, 120–25, 127–29
Coal Creek, 123
Coconino Sandstone, xi, 23, 24, 33, 41, 90, 107, 108
Colorado National Monument, 63, 159
Colorado Plateau, vii–xii, 16, 19, 21, 22, 48, 50, 53, 57, 58, 74, 79, 84, 87, 91, 103, 112, 113, 122, 124, 126–28, 135–37, 139, 140, 143, 146, 148, 150, 151, 153, 155–59, 161–63, 165, 166, 168, 170–72, 174–76, 179–82, 184, 186, 188–93, 195, 199, 200, 204, 206, 209, 211, 214, 215, 217, 218, 220, 221
Colorado River, vii, 3, 19, 24, 28, 30–33, 37, 39, 45, 54, 56, 66, 70–72, 79, 81, 145, 146, 154, 195, 198, 199, 204, 210, 214–17
Comb Ridge, 79, 146
Comb Wash, 99
Convection cells, 190, 191

Coral, 17, 18
Cordilleran Sea, ix, 13, 15, 16
Cow Springs Sandstone, 153, 156
Cretaceous Period, 161, 173, 174, 184
Crinoid, 17, 51–53, 77, 78, 123, 125, 126, 224
Cross Canyon, 67
Cunningham Gulch, 124, 125
Curtis Sandstone, 153–55
Cutler arkose, 75, 78, 95
Cutler Creek, 131
Cutler Formation, 71, 75, 76, 131
Cutler Group, 76, 79, 85, 225

Dakota Sandstone, 158, 162, 163, 165, 170
Dark Canyon, 99
Datil volcanic field, 188
Dead Horse Point, 81, 83, 220
De Chelly Sandstone, 103
Defiance Uplift, 101–3, 105, 106, 109
Devonian Period, 15, 16
Diastem, 19, 22
Dike, 42, 184, 186, 225
Dineh, 211–13
Dinosaur National Monument, 159
Dinosaurs, 42, 96, 146, 156, 158–60, 174
Diplodocus, 160
Dirty Devil River, 31
Dolomite, 13, 14, 16, 18, 36, 37, 48–51, 53, 83–85, 89, 91, 106, 109, 121–24, 128, 129, 154, 225
Dolores Formation, 141
Dolores River, 66
Dox Formation, 41
Dutton, 197

East Kaibab monocline, 37
Echo Cliffs, 32, 45
El Morro National Monument, 204
Elaterite Basin, 79, 83, 84
Elbert Creek, 122
Electra Lake, 119
Elephant Canyon Formation, 23, 72, 77, 90, 96

Kaibab Formation, 24, 25, 28, 34, 84, 85
Kaibab Limestone, 37, 41, 109
Kaibab Trail, 15, 43
Kaibab uplift, 15, 21, 25, 28, 37, 40, 92, 101, 136, 165, 195, 198, 199
Kayenta Formation, 145, 146

La Plata Mountains, 139
La Sal Mountains, 158, 189
Lake Bidahochi, 199
Lake Mead, 30, 31, 39, 188, 198
Lake Powell, 30, 39, 94, 145, 214
Laramide orogeny, 176, 179, 182, 188, 192, 226
Lateritic soil, 21, 126
Leadville Formation, 124
Lee's Ferry, 31
Lewis Shale, 172
Lime Ridge anticline, 93, 99
Limestone, 4, 7, 9, 13, 16–19, 21, 22, 24, 25, 32–35, 41, 48–51, 53, 57–60, 63, 70, 71, 76, 77, 83, 84, 89–91, 94–96, 106, 108, 109, 111, 119, 123, 124, 126, 127, 129–31, 137, 140, 147, 150, 151, 154, 156, 224–27
Lisbon Valley, 49, 51, 158, 220
Little Colorado River, 15, 24, 25, 39, 40, 136, 198, 199

Mancos Shale, vii, 165, 166, 170, 172, 188
Manuelito, 212
Marble Canyon, 15, 31, 32, 37, 39, 45, 47, 79, 198
Marble Canyon Damsite, 36
Matterhorn, 202
Mazatzal uplift, 109
McDermott Formation, 173, 184
McIntyre Canyon field, 220
Mendocino lineament, ix, 192
Menefee Formation, 168, 210
Mesa Verde National Park, 165, 166, 168, 170, 210
Mesaverde Group, 166, 168, 170, 172, 173, 176
Meseta Blanca Member, 106

Mesozoic Era, 45, 91, 174
Metamorphic rocks, 5–7, 40, 47, 62, 63, 14, 115, 119
Mill Creek, 125
Mississippian Period, 16, 18, 19
Mitten Butte anticline, 94
Moab, 49, 58, 66, 71, 75, 76, 81, 126, 146, 151–53, 155, 166, 180, 189, 215, 220
Moenave Formation, 146
Moenkopi Formation, 84, 101, 136, 137, 140
Molas Formation, 57, 58, 126, 127
Molas Lake, 57, 117, 121–27, 129
Molas Pass, 57, 123, 129, 203
Mollusks, 25, 36, 154, 174, 224
Monument Upwarp, 45, 70, 78, 79, 92–95, 97, 99–101, 136, 148, 151, 195, 199, 219
Monument Valley, vii, 23, 76, 79, 95, 96, 100, 101, 106, 139, 148, 219
Moraines, 202, 203
Morrison Formation, 150, 156–59, 161, 220
Mount Taylor, 204
Muav Limestone, 13, 14, 36, 37, 48, 89
Museum of Northern Arizona, 198

Nacimiento Mountains, 111
Nankoweap Canyon, 37
Narbona, 212
Natural Bridges National Monument, vii, 99, 137
Navajo Bridge, 32, 33, 45
Navajo Mountain, 148
Navajo Sandstone, xi, 145–48, 150
Needles District, 69, 70
Needles Mountains, 115, 117, 118
North Horn Formation, 173

Organ Rock Shale, 79, 100, 148
Orogeny, 175, 176, 179, 182, 188, 192, 226
Ostracods, 161
Ouray Limestone, 49, 50, 123, 125

Triassic Period, 111
Trilobites, 14, 25, 37, 42, 121
Twin Creek Formation, 150

Uinta Mountains, ix, 79, 154, 155, 181, 184, 195, 217
Uncompahgre Canyon, 117, 118, 124, 127
Uncompahgre Creek, 123
Uncompahgre Plateau, 54, 62, 63, 148, 156
Uncompahgre Quartzite, 117, 121, 122
Uncompahgre uplift, viii, 54, 60, 63, 75, 76, 85, 95, 100, 101, 103, 126, 127, 131, 141, 151, 159
Unconformity, 6, 7, 15, 19, 21, 22, 24, 25, 34, 37, 40, 41, 71, 119, 131, 135, 143, 148, 150, 153, 156, 157, 162, 163, 176, 223, 225, 227
Uniformitarianism, 10, 47
University of Utah, xii, 159
Unkar Rapid, 41
Upheaval Dome, 67–69
Uraninite, 158
Uranium, 135, 137, 139, 158, 177, 190, 211, 218, 220
Uravan Mineral Belt, 158, 190, 220

Vasey's Paradise, 36
Vishnu Schist, 5–7, 14, 43, 114, 117
Volcanoes, 158, 193

Walnut Canyon National Monument, 25
Wanakah Formation, 156
Wasatch Range, 137
Watahomigi Formation, 22
Waterpocket Fold, 146, 148
Westwater Canyon, 157, 158
Westwater Canyon Sandstone Member, 157
White Canyon, 99, 137, 139, 219
White House Ruin Trail, 103
White Mountains, 188
White Rim Sandstone, 80, 81, 83, 84, 90, 91, 101

Wingate Sandstone, 140, 143, 144, 148, 150
Winsor Sandstone, 156
Wrench faults, 117, 123, 180, 181

Yeso Formation, 106, 107

Zion Canyon, 150
Zuni Mountains, 105–9, 111, 204
Zuni uplift, 105, 106, 136, 140, 204